AF577239

# ULTRASTRUCTURE PROCESSING OF ADVANCED STRUCTURAL AND ELECTRONIC MATERIALS

# ULTRASTRUCTURE PROCESSING OF ADVANCED STRUCTURAL AND ELECTRONIC MATERIALS

Edited by

**L.L. Hench**

Ceramics Division
Department of Materials Science and Engineering
University of Florida
Gainesville, Florida

np

NOYES PUBLICATIONS
Park Ridge, New Jersey, USA

Library of Congress Catalog Card Number: 84-14835
ISBN: 0-8155-1004-7
Printed in the United States

Published in the United States of America by
Noyes Publications
Mill Road, Park Ridge, New Jersey 07656

10 9 8 7 6 5 4 3 2 1

Library of Congress Cataloging in Publication Data
Main entry under title:

Ultrastructure processing of advanced structural
and electronic materials.

Bibliography: p.
Includes index.
1. Ceramics. 2. Glass. 3. Glass--ceramics.
4. Composite materials. I. Hench, L.L.
TP858.U48 1984 666 84-14835
ISBN 0-8155-1004-7

# Foreword

This book presents studies in the science of ultrastructure processing. As used here, ultrastructure processing means the manipulation and control of surfaces and interfaces to obtain new, high performance materials with predictable properties and environmental insensitivity.

Materials areas that may benefit from the use of ultrastructure processing include: particulate solids behavior, adhesion of fillers and reinforcers in composites, corrosion of glasses and glass-ceramics, fatigue of brittle materials, grain boundary attack of ceramics, effects of energetic particle beams, lifetime of non-oxide ceramics, electronic behavior of high band gap semiconductors, and multiphase electronic components.

Approaches to ultrastructure processing described in the book are: (1) a systematic investigation of environment-surface interactions of silicate glasses and glass-ceramics, and non-oxide ceramics; (2) the production of unique ceramics, glasses, glass-ceramics, and composites by the use of "transformation processing" (producing materials directly from chemical conversions rather than traditionally forming materials by compacting and densifying large particulates into objects); and (3) "micromorphology processing"—production of submicron spherical powders, control of powder surface chemistry, and controlled assembly of particulates into ceramic or composite bodies.

The information in the book is from the following documents:

> *Ultrastructure Processing and Environmental Stability of Advanced Structural and Electronic Materials—Final Technical Report,* by L.L. Hench of University of Florida, Department of Materials Science and Engineering for the U.S. Air Force Office of Scientific Research, April 1983.

*Ultrastructure Processing and Environmental Stability of Advanced Structural and Electronic Materials—Third Annual Report,* edited by L.L. Hench of University of Florida, Department of Materials Science and Engineering for the U.S. Air Force Office of Scientific Research, March 1983.

The table of contents is organized in such a way as to serve as a subject index and provides easy access to the information contained in the book.

Advanced composition and production methods developed by Noyes Publications are employed to bring this durably bound book to you in a minimum of time. Special techniques are used to close the gap between "manuscript" and "completed book." In order to keep the price of the book to a reasonable level, it has been partially reproduced by photo-offset directly from the original reports and the cost saving passed on to the reader. Due to this method of publishing, certain portions of the book may be less legible than desired.

## NOTICE

# Contents and Subject Index

# Part I
# Overview

The information in Part I is from *Ultrastructure Processing and Environmental Stability of Advanced Structural and Electronic Materials—Final Technical Report,* by L.L. Hench of University of Florida, Department of Materials Science and Engineering for the U.S. Air Force Office of Scientific Research, April 1983.

# SUMMARY AND OVERVIEW OF MULTI-INVESTIGATOR RESEARCH PROGRAM (MIRP)

The goal of our Multi-Investigator Research Program (MIRP) is to achieve an understanding of the science of ultrastructure processing of ceramics, glasses and composites. Ultrastructure processing as used in our program and in this report refers to the manipulation and control of surfaces and interfaces for the purpose of attaining a new generation of high performance materials with predictable properties and environmental insensitivity or control.

There are three primary chemistry based approaches that we are taking to control surfaces and interfaces in a step by step method (Fig. 1). Our first approach is a systematic investigation of environment-surface interactions of silicate glasses and glass-ceramics, and non-oxide ceramics. The second major thrust of the MIRP is the production of unique ceramics, glasses, glass-ceramics, and composites by use of what we term "transformation processing". It is essentially making materials directly from chemical conversions in contrast to traditional methods involving large particulates which are compacted and densified into objects. The third ultrastructure processing area of the MIRP is what we term "micromorphology processing". There are three aspects to the concept of micromorphology processing: 1) production of submicron spherical powders, 2) control of powder surface chemistry, and 3) controlled assemblage of the particulates into either ceramic or composite bodies. By manipulation of all three aspects of micromorphology processing it should be possible to tailor-make the microstructure, grain boundaries,

ULTRASTRUCTURAL PROCESSING OF HIGH PERFORMANCE CERAMICS

APPROACH

IMPROVED PERFORMANCE THROUGH CONTROLLED PROCESSING

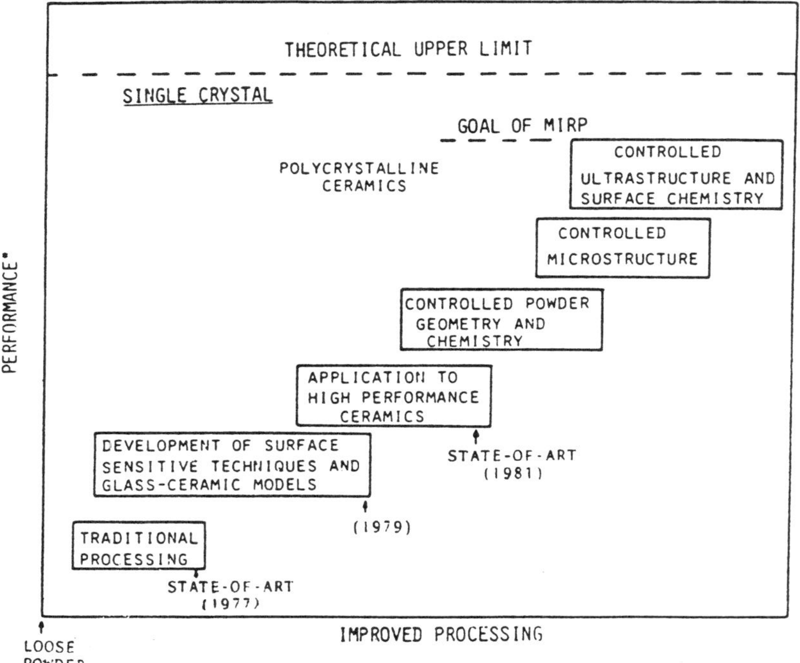


*MECHANICAL STRENGTH, CHEMICAL RESISTANCE, ENVIRONMENTAL STABILITY

Figure 1.

interfaces, and surfaces of ceramics and composites to meet unique combinations of property-environment requirements.

By understanding the deterioration mechanisms of several classes of materials we are establishing how processing must be controlled to eliminate or minimize the impact of environmental deterioration. Simple, binary alkali silicate glasses are the model systems for these studies. Aqueous corrosion of these glasses is now reasonably well understood and the corrosion mechanisms are applicable to a very wide range of silicate glass compositions and types of environmental exposures, as described in a series of papers listed herein. We have introduced the concept of type III glass surfaces, possessing dual or multiple protective surface films and have identified many features of such materials and the mechanisms of their development.

Details of the mechanisms of atmospheric weathering of glasses are less known than for aqueous corrosion. Consequently, an effort in our program for several years was to develop models for atmospheric weathering. We have extended this work to include hydrothermal reactions of glasses and two phase glass-ceramics as well. Comparison of aqueous hydrothermal corrosion of two phase lithia-silicate glasses and glass-ceramics and hydrothermal vapor attack of the same materials shows enhanced resistance to vapor phase attack due to surface crystallization of glassy phases. We have shown that stability of the glass-ceramics improves with increases in the volume fraction of crystal phase. Interpretation of the corrosion mechanisms of the two phase glass-ceramics was made possible by the development of polarized infrared reflection spectroscopy (PIRRS) for surface analysis of polcrystalline ceramics and glass-ceramics during the contract. The PIRRS technique was applied to the problem of aqueous corrosion of lithia disilicate glass-ceramics.

Understanding the influence of environmental exposure on the mechanical behavior of glasses and ceramics is essential if performance lifetimes are to be improved. We have used 33L lithia silica glasses and glass-ceramics as a model system to attack this difficult problem. Work reported during several years of this program has determined the effect of the distributed glassy phase in aqueous corrosion, weathering, and hydrothermal attack of the materials. The same materials have also been tested using both dynamic fatigue and crack growth methods to determine their environmental sensitivity to lifetime predictions. This effort shows that stress corrosion susceptibility is more sensitive on aging in water than on percent of crystallinity. This is due to the importance of interphase attack in two phase systems. Important differences occur in the environmental sensitivity determined by dynamic fatigue vs crack growth methods leading to major differences in lifetime predictions for the same materials. This finding shows that materials with a reactive secondary phase need to be tested with more than one method to make reliable lifetime predictions.

Use of electron beam techniques is essential for understanding surface layers formed on glasses, ceramics, and composites. Thus understanding the nature of the electron beam interaction with a glass surface is critical for achieving an accurate chemical analysis. Also, impingement of electrons on a glass surface is an important class of environmental interaction in its own right. Several years of experiments in this area have culiminated in a general model of alkali desorption and diffusion during electron bombardment of glass. The model is based upon experiments that show alkali ions being converted to a new activated state within the solid by interaction with the incident electron beam. Either diffusion or desorption of the altered alkali ions can follow this process. This concept of an electron-induced surface complex satisfactorily explains both incubation times and Auger electron signal decay processes observed for glasses. Our work in this area has also shown that $Na^+$, $O^+$ and $Si^+$ desorb from soda-silicate glass surfaces and involve both inter-atomic and intra-atomic Auger de-excitations. $Si^+$ desorption occurs only after electron bombardment creates a Si-rich surface.

Because of the extremely small scale of secondary phases in SiC and $Si_3N_4$ ceramics it is difficult to understand their contribution to environmental sensitivity. Auger electron spectroscopy and Fourier transform IR reflection spectroscopy (FTIRRS) have been used to characterize surfaces and interfaces of polycrystalline and single crystal SiC. Scanning AES of polycrystalline SiC fracture surfaces formed in-situ showed that boron does not segregate to or precipitate at $\alpha$-$\beta$ phase boundaries, in contrast to many hypotheses. However, low temperature (1100°C) oxidation of polycrystalline SiC does implicate segregation of boron to the oxide.

FTIRRS analyses of the same polycrystalline SiC samples quite clearly shows evidence of formation of an intermediate Si-O-C phase forming under the surface $SiO_2$ phase during 1100°C oxidation in pure $O_2$. A critical incubation or nucleation time is required for the mixed oxycarbide phase to form. Both the structure and kinetics of this intermediate phase differ from single crystal SiC. Sequential oxidation, exposure at 1100°C followed by cooling then reheating, results in very different mechanisms than continuous oxidation.

Our efforts to relate these environmental effects to the electronic behavior of SiC are only just beginning. However, considerable progress has been made in characterizing the electron behavior of the anisotropic sub-mini bands associated with SiC polytypism boundaries. By use of anisotropic noise measurements it has been possible to quantify the nonuniform distribution of mobile carriers along the c-axis direction of a SiC single crystal. Majority carrier injection occurs at the high resistivity boundary and electron conduction mechanism in the resistive region controls current-voltage, impedance, and noise characteristics of the crystal. Quantitative comparison of these experimental results with a polytypic band theory model is well underway.

For advanced electrical applications such as batteries or fuel cells, ionic transport in glasses are as important as electrons. During the last two years we have enjoyed a collaboration between our MIRP and Prof. Abe's glass research group at Nagoya Institute of Technology. This has led to several publications, including a major study on protonic conduction in phosphate glasses. Other investigations leading to publication include photo and thermo coloring of phosphate glasses, an intriguing physical phenomena and also an important extension of the standard IR method for determining water content in glasses. The latter work is of special interest to us because of our MIRP redirection during year 03 to pursue a vigorous effort in sol-gel processing of glasses, ceramics, and composites. Materials made from such chemically based processing can possess a significant concentration of residual organic and aqueous constituents. Quantitative analysis of the kinetics of drying and gel-glass conversion is now one of our prime emphases in this area.

Our efforts in sol-gel or chemical transformation processing have been primarily concentrated on alkali-silicate glass systems. We have established some general principles regarding the compositional dependence of $Na_2O$-$SiO_2$ gel formation and their environmental sensitivity. Thermal processing diagrams have been developed for these compositions as well. Of particular importance to eventual applications is the discovery of a broad compositional range where very low temperature (460-520°C) full densification can be obtained.

Our second direction in transformation processing has been directed towards development of SiC or $Si_3N_4$ ceramics from organometallic percursors. Three different methods are being explored. RF plasma polymerization has been used to produce thin, comformable, crosslinked polymer-coatings up to 1 μm thick. Upon pyrolysis the coatings are converted to ceramic-like SiC or $Si_3N_4$ layers. Such coatings exhibit superb corrosion protection of steels.

A second method of forming SiC involves pyrolysis of polysilastyrene polymers, produced by Prof. West of the University of Wisconsin. Extrusion of fibers and pyrolysis to a SiC-like ceramic has been demonstrated. The third method is a chemical vapor deposition process for forming uniform, spherical 0.2 to 10 μm powders of SiC. Patent applications are being prepared for the RF plasma and CVD processes.

The third major emphasis of the MIRP is what we term "micromorphology processing". During the last several years we have made progress in the preparation of submicron spherical powders of SiC and $Al_2O_3$ and $SiO_2$. In order to manipulate such powders into controlled assemblages though requires a complete understanding of their surface chemistry. A major development has been achieved in the theoretical framework for understanding electrochemical reactions at oxide-water interfaces. This new theory involves a variable point of zero charge model and incorporates two surface charging mechanisms, one fast and one slow. These modifications make it possible to understand aging and dehydration effects as well as the influence of various electrolyte on surface potentials. It may be possible with this theoretical basis to understand the consequences of aging of gel-derived glasses and ceramics and the effects of prolonged aqueous exposure on lifetimes of materials. A unique aspect of our studies of micromorphology processing is the combined use of micromechanics and surface chemistry to achieve control over compaction and forming of ceramics. Two studies have provided some of the scientific foundation for this subject. They provide the conceptual frame work for using the submicron spherical powders with tailored surfaces to generate unique microstructures and defect free compacts. The use of graphical analysis of specific volume diagrams can provide the necessary reference points for studying the evolution of microstructure during firing, identify the relative effects of solid-liquid ratios on density, and analyze bulk and apparent specific volumes of polymer-ceramic composites.

An eventual goal of the MIRP is to design controlled ceramic or glass filled polymer or co-polymer systems with chemically bonded interfaces. The results of a study of the compatibility range of random copolymer of varying composition with each homopolymer, especially poly(styrene), poly(n-butylmethacrylate), and poly(styrene-co-n-butylmethacrylate) have been reported.

Our studies of co-polymer composites have demonstrated that it is possible to modify the co-polymer chemistry in such a way as to control the interface between glass and polymer. Large variations in elastic modulus and strength of the composite system result from the chemical manipulation of the interfaces within the composite. The next step will be manipulation of the surfaces of the glasses by adsorbing ions which will interact with charge groups on the co-polymer. Incorporation of the spherical silica, alumina, SiC, and $Si_3N_4$ powders into the co-polymer composites with tailored surfaces is our long range objective.

## LIST OF PUBLICATIONS AND PRESENTATIONS EITHER PARTIALLY OR FULLY SUPPORTED BY AFOSR CONTRACT

1. "Surface Characterization of Ceramed Composites and Environmental Sensitivity," presented at the Conf. on Composites and Advanced Materials, Cocoa Beach, Florida, January 1978.

2. "Aqueous Corrosion of Lithia-Alumina-Silicate Glasses," Am. Ceram. Soc. Bull. 57[11] 1040-1044 (1978).

3. "Effects of Glass Surface Area to Solution Volume Ratio on Glass Corrosion," J. Phys. & Chem. Glasses 20[2] 35-40 (1979).

4. "Ceramic Surface Degradation," presented at 70th Annual Meeting of American Institute of Chemical Engineers, New York City, Nov. 1977.

5. "Physical Chemistry of Glass Surfaces," J. Non-Crystalline Solids 28, 83-105 (1978).

6. "Physical Chemistry of Glass Surfaces," Survey Lecture for the XIth International Congress of Glass Conf., Prague, Czechoslovakia, July 4, 1977.

7. "Physical Chemistry of Glass Surfaces," Glass '77, A Survey of Contemporary Glass Science and Technology, Jeri Gotz, ed., CVTS-DUM Techniky, Praha, Czechoslovakia, 343-370, July 1977.

8. "Surface Analysis of $Si_3N_4$ Oxidation," presented at the Conf. on Composites and Advanced Materials, Cocoa Beach, Florida, January 1978.

9. "Corrosion Behavior of Lithia Disilicate Glass in Aqueous Solutions of Aluminum Compounds," Am. Ceram. Soc. Bull. 58[11] 1111 (1979).

10. "Determination of Fatigue Parameter for $Li_2O \cdot 2SiO_2$ Glass and Partially Crystallized $Li_2O \cdot 2SiO_2$," presented at the 80th Annual Meeting of the American Ceramic Society, Detroit, Michigan, May 6-11, 1978.

11. "Effect of Crystallization on the Auger Electron Signal Decay in a $Li_2O \cdot 2SiO_2$ Glass and Glass-Ceramic," J. Am. Ceram. Soc. 62[9-10] 500-503 (1979).

12. "Durability of $Li_2O \cdot 2SiO_2$ Glass-Ceramics," presented at the 80th Annual Meeting of the American Ceramic Society, Detroit, Michigan, May 6-11, 1978.

13. "Compositional Analysis of Alkali Glasses Using Auger Electron Spectroscopy," presented a the 80th Annual Meeting of the American Ceramic Society, Detroit, Michigan, May 6-11, 1978.

14. "Fatigue Properties of $Li_2O \cdot 2SiO_2$ Glass and Glass-Ceramics," J. Am. Ceram. Soc. 62[5-6] 319-320 (1979).

15. "Corrosion of Glass Surfaces," Surface Science 100, 53-70 (1980).

16. "Use of New Surface Physics for Controlling the Physical Properties of Ceramics," presented at the Science of Ceramics 10, Munich, Germany, Sept. 2-4, 1979.

17. "Use of New Surface Physics for Controlling the Physical Properties of Ceramics," Proceedings Science of Ceramics 10, 669-680 (1980).

18. "Surface Characterization of Ceramed Composites and Environmental Sensitivity," in Ceramic Engineering and Science Proceedings, Vol. 1, No. 7-8(A), pp. 311-318, The American Ceramic Society, Columbus, Ohio (1980).

19. "Infrared Reflection Analysis of $Si_3N_4$ Oxidation," in Ceramic Engineering and Science Proceedings, Vol. 1, No. 7-8(A), pp. 318-331, The American Ceramic Society, Columbus, Ohio (1980).

20. "Effects of Heat Treatment Time on the Oxidation of Pressed $Si_3N_4$ as Determined by Infrared Reflection Analysis," in Ceramic Engineering and Science Proceedings, Vol. 1, No. 7-8(A), pp. 481-488, The American Ceramic Society, Columbus, Ohio (1980).

21. "Durability and Strength of $Li_2O$-$Al_2O_3$-CaO-$SiO_2$ Glass-Ceramics," presented at 81st Annual Meeting of the American Ceramic Society, Cincinnati, Ohio, April 29-May 2, 1979.

22. "Effects of Solution pH on Corrosion Behavior of $Li_2O \cdot 2SiO_2$-Glass-Ceramics," presented at the 81st Annual Meeting of the American Ceramic Society, Cincinnati, Ohio, April 29-May 2, 1979.

23. "Nondestructive Evaluation of Surface Flaws Using Infrared Reflection Spectroscopy," presented at 81st Annual Meeting of the American Ceramic Society, Cincinnati, Ohio, April 29-May 2, 1979.

24. "Corrosion of Glass Surfaces," an invited paper presented at the Seventh Canadian Seminar on Surfaces, Pinawa, Canada, July 1979.

25. "Electron Beam Effects During Analysis of Glass Thin Film with Auger Electron Spectroscopy," J. of Surface and Interface Analysis 2, 85 (1980).

26. "Novel Techniques for Formation of Silicon Carbide," presented at the 5th Annual Conf. on Composites and Advanced Materials, Merritt Island, Florida, January 18-22, 1981.

27. "T.E.M. Application to Aspects of Ceramic Processing," presented at the 5th Annual Conf. on Composites and Advanced Materials Merritt Island, Florida, January 18-22, 1981.

28. "Future Direction for Ceramic Processing Science," presented at the 5th Annual Conf. on Composites and Advanced Materials, Merritt Island, Florida, January 18-22, 1981.

29. "Nuclear Wastes--A Problem Requiring Innovations in Processing and Advanced Materials," presented at the 5th Annual Conf. on Composites and Advanced Materials, Merritt Island, Florida, January 18-22, 1981.

30. "Weathering of $Na_2O$ $2SiO_2$· Glass," presented at the American Ceramic Society Glass Division Fall Meeting, Bedford Springs, Pennsylvania, Oct. 8-10, 1980.

31. "Control of Glass and Glass Surfaces - The 1980 George W. Morey Award Lecture," presented at the American Ceramic Society Glass Division Fall Meeting, Bedford Springs, Pennsylvania, Oct. 8-10, 1980.

32. "Surface Physics of Glass," presented at the Swedish Dept. of Defense Laboratories, Stockholm, Sweden, Sept. 1980.

33. "Electron Stimulation Desorption from Soda-Silica Glass Surface," presented at the 9th Users Conf./SIMS, Minnesota, June 4-16, 1980.

34. "Electron Stimulation Desorption from Soda-Silica Glass Surface," presented at the American Vacuum Society Meeting, Tampa, Florida, Feb. 11, 1981.

35. "Application of Rutherford Backscattering Technique to Glass Surface Analysis," presented at the 3rd Appl. Surf. Sci. Symp., Dayton, Ohio, June 3-6, 1980.

36. "General Model of Sodium Desorption and Diffusion During Electron Bombardment of Glass," J. Vac. Sci. Technol., in press.

37. "Surface and Interface Characterization of Advanced Materials," presented at the 6th Annual Conf. on Composites and Advanced Ceramic Materials, Cocoa Beach, Florida, Jan. 17-21, 1982.

38. "Preferential Sputtering of PtSi, $NiSi_2$, and AgAu," J. Vac. Sci. Technol., in press.

39. "Correction Factors and Preferred Sputtering in Quantitative Auger Electron Spectroscopy," Surf. Interface. and Anal., submitted.

40. "General Model of Sodium Desorption and Diffusion During Electron Bombardment of Glass," presented at the 28th National Symp. Amer. Vac. Soc., Anaheim, California, Nov. 2-6, 1981.

41. "General Model of Sodium Desorption and Diffusion During Electron Bombardment of Glass," invited presentation Physical Electronics Div., Perkin Elmer Corp., August, 1981.

42. "Preferential Sputtering of PtSi, $NiSi_2$, and AgAu," presented at the 28th National Symp. Amer. Vac. Soc., Anaheim, California, Nov. 2-6, 1981.

43. "Quantitative Surface Analysis," invited presentation National Phys. Lab., Traddington, England, Nov. 24-25, 1981.

44. "General Model of Sodium Desorption and Diffusion During Electron Bombardment of Glass," invited presentation Tektronix, Inc., Beaverton, Oregon, March 4, 1982.

45. "Weathering of Binary Alkali Silicate Glasses and Glass-Ceramics," presented at the 6th Annual Conf. on Composites and Advanced Ceramic Materials, Cocoa Beach, Florida, Jan. 17-21, 1982.

46. "Surface Behavior of Gel-Derived Glasses," presented at the 6th Annual Conf. on Composites and Advanced Ceramic Materials, Cocoa Beach, Florida, Jan. 17-21, 1982.

47. "Low Temperature Surface Reactions of SiC," presented at the 6th Annual Conf. on Composites and Advanced Ceramic Materials, Cocoa Beach, Florida, Jan. 17-21, 1982.

48. "Preparation of 33 Mole % $Na_2O$ - 67 Mole % $SiO_2$ Glass by Gel-Glass Transformation," J. Non-Crystalline Solids 53, 183-193 (1982).

49. "Novel Techniques for the Formation of Silicon Carbide," seminars presented March 1981 at Dow-Corning Corp. and April 1981 at Eastman Kodak Corp.

50. "Compatability of Polystyrene and Poly(styrene-co-n-butylmethacrylate) Blends," ACS Meeting, New York, August 1981; Organic Coatings and Plastics Preprints 44[2] 73 (1981).

51. "Formation of Silicon Carbide Powders by Vapor Phase Reactions," presented at the 6th Annual Conf. on Composites and Advanced Ceramic Materials, Cocoa Beach, Florida, Jan. 17-21, 1982.

52. "Plasma Polymerization of Silanes," presented at ACS Meeting, Las Vegas, March 1982; Organic Coatings and Applied Polymer Science Preprints 45[1] 127 (1982).

53. "Composites of Glass Particles in Poly(styrene-co-n-butylmethacrylate) and Modified Copolymers," presented at ACS Meeting, Las Vegas, March 1982; Organic Coatings and Applied Polymer Science Preprints 46[1] 115 (1982).

54. "Modification of Poly(styrene-co-n-butylmethacrylate) Copolymers," presented at ACS Meeting, Las Vegas, March 1982; Organic Coatings and Applied Polymer Science Preprints 46[1] 440 (1982).

55. "XPS Study of Plasma Deposited Films Containing Silicon," presented at ACS Meeting, Las Vegas, March 1982; Organic Coatings and Applied Polymer Science Preprints 46[1] 134 (1982).

56. "Treatment of Glass Surfaces for Polymer-Glass Composites," presented at the 7th Annual Conf. on Composites and Advanced Ceramic Materials, Cocoa Beach, Florida, Jan. 16-19, 1983.

57. "Phase Transformations in Sol-Gel Derived Aluminas," presented at the International Conference on Ultrastructure Processing of Ceramics, Glasses and Composites, Feb. 13-17, 1983, Gainesville, Florida, L. L. Hench and D. R. Ulrich, eds., J. Wiley & Sons (1983).

58. "Static Aqueous Corrosion of Soda-Disilicate Glass (33N) in Simulated Geologic Environments," presented at Annual Meeting of the Amer. Ceram. Soc., Cincinnati, Ohio, May 5, 1983.

59. "Weathering of Binary Alkali Silicate Glasses and Glass-Ceramics," in Ceramic Engineering ans Science Proceedings, 3[9-10 458-476 (1982).

60. "Surface Analysis of Glasses," in Industrial Applications of Surface Analysis, L. A. Casper and C. J. Powell, eds., ACS Symp. Series, ISSN 009-6156, 199 (1982) pp. 203-229.

61. "Surface Behavior of Gel-Derived Glasses," in Ceramic Engineering and Science Proceedings, 3[9-10] 477-483 (1982).

62. "Low Temperature Oxidation of SiC," in Ceramic Science and Engineering Proceedings, 3[9-10] 596-600 (1982).

63. "Glass Surfaces - 1982," in Proceedings of International Conf. on Physics of Amorphous Solids, J. Zarzycki, eds., Montpellier, France, July 1982.

64. "Preparation of $xNa_2O-(1-x)SiO_2$ Gels for the Gel-Glass Process. Part I: Atmospheric Effect on the Structural Evolution of the Gels," J. Non-Cryst. Solids 48, 79-95 (1982).

65. "Preparation of $xNa_2O-(1-x)SiO_2$ Gels for the Gel-Glass Process. Part II: The Gel-Glass Conversion," J. Non-Cryst. Solids (in press).

66. "Physical Chemical Factors in Sol-Gel Processing," presented at the International Conference on Ultrastructure Processing of Ceramics, Glasses and Composites, Feb. 13-17, 1983, Gainesville, Florida, L. L. Hench and D. R. Ulrich, eds., J. Wiley & Sons (1983).

67. "Processing and Environmental Behavior of a 20 Mol % $Na_2O$ - 80 Mol % (20N) $SiO_2$ Gel Glass," presented at the 7th Annual Conf. on Composites and Advanced Ceramic Materials, Cocoa Beach, Florida, Jan. 16-19, 1983.

68. "Aqueous Durability of Lithium Disilicate Glass-Ceramics," Bull. Am. Ceram. Soc. 61[11] 1218-1223 (1982).

69. "Physical Parameters for Characterizing Agglomerates," presented at the Annual Meeting of the American Ceramic Society, Cincinnati, OH, May 5, 1982.

70. "Role of Effective Stress in Extrusion," presented at the Annual Meeting of the American Ceramic Society, Cincinnati, OH, May 5, 1982.

71. "Application of Specific Volume Diagrams to Ceramic Processing," presented at the Annual Meeting of the American Ceramic Society, Cincinnati, OH, May 5, 1982.

72. "Surface Chemistry of Oxides in Water," presented at the International Conference on Ultrastructure Processing of Ceramics, Glasses and Composites, Feb. 13-17, 1983, Gainesville, Florida, L. L. Hench and D. R. Ulrich, eds., J. Wiley & Sons (1983).

73. "Application of Solid Mechanics Concepts to Ceramic Particulate Processing," in Advances in Powder Technology, G. Chin, eds., Am. Soc. for Metals, 1982.

74. "Specific Volume Diagrams for Ceramic Processing," J. Am. Ceram. Soc. 66[4] 297-301 (1983).

75. "Surface Chemistry in Water," presented at the International Conference on Ultrastructure Processing of Ceramics, Glasses and Composites, Feb. 13-17, 1983, Gainesville, Florida, L. L. Hench and D. R. Ulrich, eds., J. Wiley & Sons (1983).

76. "$Si_3N_4$ and SiC from Organometallic Precursors," presented at the International Conference on Ultrastructure Processing of Ceramics, Glasses and Composites, Feb. 13-17, 1983, Gainesville, Florida, L. L. Hench and D. R. Ulrich, eds., J. Wiley & Sons (1983).

77. "Compatibility of a Random Copolymer with Each Homopolymer," presented at the ACS/AIChE joint meeting on Polymer Compatibility, Nov. 14, 1982, submitted for publication in monograph of meeting.

78. "Surface Structure and Properties of R.F. Plasma Polymerized Hexamethyl Disilazane," Organic Coatings and Applied Polymer Sci., 47, Sept. 1982.

79. "Compatibility of Polystyrene-Polystyrene-co-n-butyl methacrylate, Poly(n-butyl methacrylate)-Poly(styrene-co-n-butyl methacrylate) and Poly(styrene-co-n-buty methacrylate)-Poly(styrene-co-n-butyl methacrylate) Systems," National Technical Conf., Soc. of Plastics Engineerings, Symp. on Polymer Alloys, Blends and Composites, Oct. 25, 1982.

80. "Characterization of Polysilastyrene," presented at the High Polymer Div., Am. Phys. Soc. Meeting, Los Angeles, CA, March 21-25, 1983.

81. "Pyrolysis of Polysilastyrene Thin Films as Studied by XPS and IR," presented at the High Polymer Div., Am. Phys. Soc. Meeting, Los Angeles, CA, March 21-25, 1983.

82. "Integranular Segregation of Boron in Sintered Silicon Carbide," submitted to J. Amer. Ceram. Soc., 1983.

83. "Mechanisms of Electron Stimulated Desorption from Soda-Silica Glass Surfaces," submitted to Surf. Interface Anal., 1983.

84. "Electron Beam Effects During Analysis of Glass Surfaces," presented at Max Planck Inst. fur Metallsforschung, Stuttgart, Germany, May 24, 1982.

85. "Mechanism of ESD from Glass Surfaces, presented at Florida Chapter of Am. Vacuum Soc., 12th Annual Symp., Clearwater Beach, FL, Feb. 16, 1983.

86. "Electron Stimulated Desorption: Application and Mechanisms," Pittsburgh Conf. on Analytical Chemistry and Applied Spectroscopy, Atlantic City, NJ, March 10, 1983.

87. "Electron Beam Effects During Glass Analysis," National Bureau of Standards, Gaithersburgh, MD, March 7, 1983.

# Part II

# Ultrastructure Processing Technology Studies

The information in Part II is from *Ultrastructure Processing and Environmental Stability of Advanced Structural and Electronic Materials—Third Annual Report,* by L.L. Hench of University of Florida, Department of Materials Science and Engineering for the U.S. Air Force Office of Scientific Research, March 1983.

# GLASS SURFACES—1982

L.L. Hench

*Department of Materials Science and Engineering*
*University of Florida*
*Gainesville, FL 32611*

Abstract

A new category of glass surface is described. Type IIIB surfaces are composed of multiple layers of oxides, hydroxides, and hydrated silicates resulting from a sequence of solution-precipitation reactions between the glass surface and leaching solutions. Very low leach rates and then ion depletion depths of certain compositions of complex nuclear waste glasses are due to Type IIIB surfaces. The solubility limits that establish the equilibrium ionic concentrations for burial ground waters also establish the multiple barrier Type IIIB films, to protect nuclear waste glasses in contact with those ground waters.

## Introduction

Understanding and controlling the surface reactions of glass is a subject of major technical importance. It has long been recognized that glass performance, such as the durability of glass containers, the useful life of windows and optical glasses, pH electrode lifetime, and static fatigue is related to glass surface - environment interactions. During the last five years there has been a growing awareness that the reliability of glasses developed for immobilization of high level radioactive wastes is also dependent on glass surface chemistry. We now know that the mechanisms of nuclear waste glass leaching comprise an important new category of glass surface behavior.

Understanding the kinetics and mechanisms of the new type of glass-environment interaction characteristic of nuclear waste glasses is important for several reasons. First, the quantity of nuclear waste glass required for immobilization of defense and commercial radioactive waste is millions of kilograms and the estimated cost is billions of dollars. Secondly, there must be reasonable assurance that the glass surface will be stable in contact with ground waters during geologic burial for thousands to hundreds of thousands of years. Thirdly, the mechanisms responsible for stabilization of nuclear waste glasses might be applied in the design of glasses for redox-sensing (Eh) electrodes or glasses with unique gradients in optical properties. Lastly, our ability to generalize glass-environment interactions is expanded with this new category of surface behavior.

Thus theprimary objectives of this paper are to review the evidence for a new class of glass surface behavior, compare this behavior with other categories of glass surfaces, and relate the significance of the surface structure and reaction kinetics of these glasses to the geologic performance of nuclear waste glasses.

## Background

Previous efforts to generalize the surface behavior of silicate glasses proposed five types of glass surfaces to represent a broad range of glass-environment interactions, Figure 1.[1,2] It was suggested that any silicate glass could be described in terms of one of the five surface types at any particular instant in its processing and environmental history. Recent studies indicate however that another type of surface, IIIB, must be added to this scheme. Figure 1 incorporates this change.

The ordinate in Figure 1 represents the relative concentration of $SiO_2$ (or other oxides in Type III surfaces) in the glass and the abscissa corresponds to the depth into the glass surface. If species are selectively dissolved from the glass surface the relative $SiO_2$ concentration will increase producing a $SiO_2$-rich surface layer. If all species in the glass are dissolved simultaneously (congruent dissolution) the relative concentration of

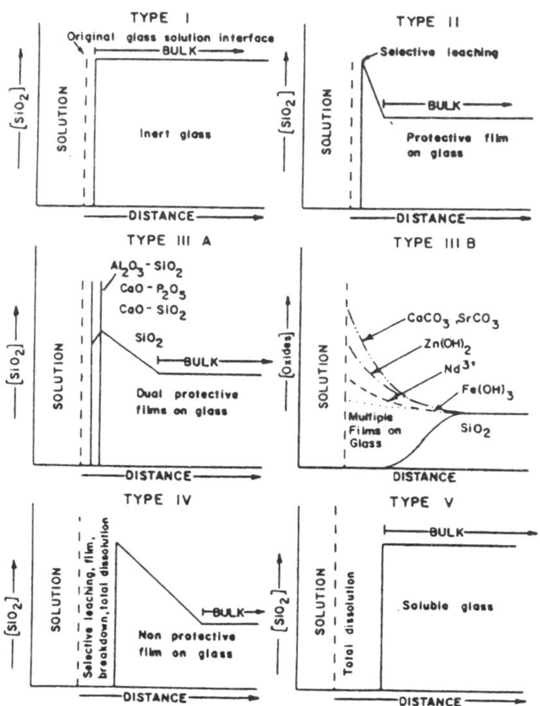


Fig. 1. Five types of glass surfaces and six surface conditions resulting from glass-environment interactions.

$SiO_2$ will remain the same as in the original glass. When combinations of selective dissolution, congruent dissolution and precipitation from solution occur then any one of the 6 surface conditions shown in Figure 1 is possible.

Type I glasses have undergone a surface reaction that is only a monolayer thick and no compositional profile is measureable. High purity vitreous silica exposed to neutral solutions is an example of a Type I surface. Exchange of alkali and alkaline earth ions with hydrogen and/or hydronium ions (i.e., selective dissolution) results in Type II glass surfaces if there is sufficient concentration of network formers in the surface film to stabilize it. Type III surfaces will be described later. If network formers are not sufficient, or if the environment is rich in $OH^-$ or other species which can break Si-O-Si network bonds, the surface layer is unstable and a Type IV surface is produced. A glass that is undergoing total network attack (also referred to as congruent dissolution) is described as having a Type V surface. Often from the perspective of average surface compostion there is little distinction between Type I and Type V surfaces. However, large quantities of ions are being lost from a Type V surface during corrosion and consequently extensive surface pitting can result due to localized heterogeneous attack. Additionally, large dimensional changes often accompany corrosion of glasses wtih Type V surfaces.

## Surface Reaction Kinetics

Studies of the kinetics of the above processes show that Type II surfaces are present when

$$C_1 = k_1 t^{1/2} \quad (1)$$

where $C_1$ is the concentration of species in solution, t is reaction time and $k_1$ is the rate constant for the diffusion controlled process. Type V surfaces correspond to a regime of kinetics where

$$C_2 = k_2 t^1 \quad (2)$$

where $C_2$ and t have the same meaning as in Equation 1 and $k_2$ is the rate constant for the interface controlled network dissolution process. Type IV surfaces result when both processes occur together; e.g.

$$C_3 = k_1 t^{1/2} + k_2 t \quad (3)$$

$$C_4 = k_1 t^{1/2} + k_2 t - k_3 t^x \quad (4)$$

Continued exposure of a glass to many chemical environments leads to progressive changes in both compositon and thickness of the surface films. Figure 2 describes such a change under static corrosion or various flow rates where the ions from the glass are permitted to build up in solution and increase the pH. Selective leaching (dissolution) behavior results in formation of a silica-rich surface film and a Type II surface . With increasing time in static conditions or slow flow rates, dissolution of the surface film network occurs due to buildup of $OH^-$ in the solution leading to a Type IV surface. Eventually at high pH's the surface film network is totally destroyed and the glass is corroding congruently (i.e., Type V behavior). Congruent dissolution proceeds until complete dissolution of the glass occurs or until the solution in contact with the glass becomes saturated with respect to the individual species. Fast flow rates prevent a pH excursion and limit the onset of congruent dissolution.

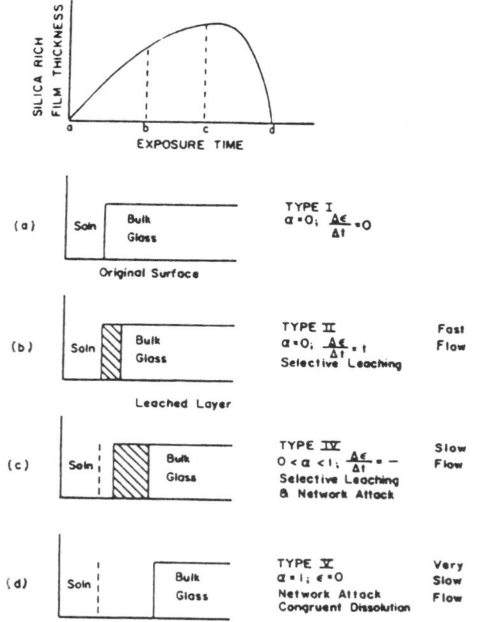


Fig. 2. Time dependent changes of glass surfaces with static leaching or flow.

A glass dissolution parameter ($\alpha$) and a film formation parameter ($\varepsilon$) can be calculated to describe these processes.[3,4] The relation of these parameters to glass surface type and reaction sequence is shown in Figure 2.

$$\alpha = \frac{\text{moles } SiO_2 \text{ in solution}}{\text{moles alkali in solution}} \Big/ \frac{\text{moles } SiO_2 \text{ in glass}}{\text{moles alkali in glass}} \qquad (5)$$

when $\alpha \to 0$, incongruent dissolution and film formation;
$\alpha \to 1$, film destruction;
$\alpha = 1$, congruent dissolution;

$$\varepsilon = \text{ppm } SiO_2 \text{(in solution)} \quad \frac{1 - \alpha}{\alpha} \qquad (6)$$

when $\dfrac{\Delta\varepsilon}{\Delta t} = +$, film deformation

when $\dfrac{\Delta\varepsilon}{\Delta t} = -$, film destruction

when $\varepsilon = 0$, congruent dissolution

The film formation parameter, $\varepsilon$, is related to the quantity of excess silica available for formation of a surface film. However, the density of the silica-rich film can vary considerably. Consequently, there may be a considerable quantity of excess silica on the glass surface of a thickness of 10-100 µm, but because of porosity and low density the film is not a barrier to diffusion (Type IV). In contrast, a thin (<1µm) film of dense $SiO_2$ remaining after selective ion exchange can serve as a highly protective surface (Type II).

A number of studies of glasses of various compositions have shown depth compositional profiles for Type II, IV, and V surfaces. The techniques used include Auger electron spectroscopy (AES)[5-8], secondary ion mass spectroscopy (SIMS)[9], resonant nuclear reactions (RNR)[10], electron spectroscopy for chemical analysis (ESCA)[11,12], secondary ion photoemission spectroscopy (SIPS)[13,14], infrared reflection spectroscopy (IRRS)[15-17]. Film depths in the range of 0.01-1.0 μm are generally observed for Type II glasses. Type IV glasses typically exhibit films of 1.0-100 μm depth. Adding network modifiers of high electric field strength to the glass changes Type IV surfaces to Type II, greatly decreases the thickness of the silica-rich films and increases the film density.

Type III Glass Surfaces

Previous surface studies of biologically active invert silicate glasses containing CaO and $P_2O_5$ (bioglasses)[18,19] and certain compositional ranges of $Li_2O$-$Al_2O_3$-$SiO_2$ glasses[20] showed that dual films developed on the glass. The secondary films were either hydrated CaO-$P_2O_5$ or $Al_2O_3$-$SiO_2$ layers. Chemical depth profiles using AES showed that the dual films formed on top of a silica rich film which resulted from rapid ion exchange of alkali for protons (or hydronium ions). Other work showed that such a dual film could be formed by addition of either phosphate[19] or aluminum ions in solution[21] as well as through release of phosphate or aluminum ions from the glass.

Analysis of anion concentrations of the dual apatite film formed on bioglasses showed some exchange of $CO_3^{2-}$ ions for $OH^-$ ions similar to that found for bone.[22] This study also showed that incorporation of $CaF_2$ in the glass results in a fluorapatite film[22] forming on the glass. Previous dental implant studies suggested a favorable tissue response to the fluoride containing glasses[23] and recent results show excellent biocompatibility of the fluoride anion doped surface layers.[24]

Studies also show that the effectiveness of a number of alkaline corrosion inhibitors[25], such as soluble calcium or beryllium salts, are due to formation of dual films composed of insoluble alkaline earth-silicate compounds[26,27].

This type of glass surface is designated type III A, dual protective film, in Fig. 1. The thickness of the secondary films can vary considerably, from as little or 0.01 μm for $Al_2O_3$-$SiO_2$ rich layers or as much as 30 μm for CaO-$P_2O_5$ rich layers.

Studies in a number of laboratories have shown that alkali borosilicate glasses designed to immobilize high level raadioactive wastes have an even more complex surface behavior than Type III A glasses. There is considerable evidence that such glasses develop multiple layers of oxides or hydroxides on their surface when exposed to water. This behavior is designated Type III B in Figure 1.

The theoretical basis for Type III B glasses is the recent investigation of B. Grambow which predicts the formation of a series of insoluble reaction products on glass surfaces.[28] He shows that the strong pH dependence of nuclear waste glass leaching can be explained in terms of the pH dependent solubility limits of simple reaction compounds of many constituents in the glass. In low and high pH leachants glass dissolution is controlled by congruent dissolution (eq. 2) but at intermediate pH values solubility of reaction species such as $Fe(OH)_3$, $Zn(OH)_2$, $Nd(OH)_3$ $SrCO_3$, $CaCO_3$, or $CaSiO_3$ control the surface reactions. Thus $k_3$ in eq. 4 involves a summation of the solution-precipitation reactions of all of the above constituents.

A mathematical model has been recently developed which attempts to describe some of the above processes.[29] It is based on the empirical observation that the concentration ratios of various elements in four different waste glass systems were independent of time in static leach tests after 3 days; however, the elemental ratios in solution differed from the corresponding ratios in the glasses throughout the 28 day static leach period. Thus, the model assumes that the glass corrodes congruently and the rate of corrosion is controlled by the rate of reaction of amorphous silica with water as well as the rate of diffusion of soluble silicates through an insoluble layer between the glass and bulk solution. This leads to the stable incongruency in the leachants. The thickness of the insoluble layer is proportional to the amount of glass that has dissolved from the start of the experiment with a silicate concentration gradient in the layer equal to the difference in silicate concentration at the glass layer interface and that in the leachrate divided by layer thickness.

Wicks and Wallace conclude,[29] "The model predicts that the slope of log concentration vs. log time plot should be initially 1 then shift to 0.5 at a later time and finally approach 0 as the leachate becomes saturated with silicates. The point at which the transition from a slope of 1 (matrix dissolution controlled) to 0.5 (diffusion controlled) occurs depends primarily on the ratio of the reaction rate constant to the diffusion coefficient. The point at which saturation becomes important depends primarily on the ratio of surface area of glass to volume of leachant. For large values of this ratio, saturation may become important before the shift from matrix dissolution to diffusion controlled kinetics occurs. In open systems where the products of leaching are not allowed to accumulate (but surface layers are formed) the slopes should shift from 1 to 0.5 but not drop below that value. A simple matrix dissolution model (involving no protective surface layers) predicts a slope of 1 which remains near that value until saturation becomes important when it then drops toward zero.

Experimental values of the slopes in four sets of static tests varied between 0.5 and 0.25. The estimated fractional saturation of silicates in these tests was sufficient to account for slopes from 0.48 to 0.41 for the surface layer model." Even lower values approaching 0.05 are observed in long term tests.[30,31]

| Oxide (%) \ Glass | M 1 | M 2 | M 3 | M 4 | M 5 | M 6 | M 7 |
|---|---|---|---|---|---|---|---|
| $SiO_2$ | 47.6 | 42.7 | 51.13 | 45.6 | 50.0 | 48.4 | 46.1 |
| $Al_2O_3$ | - | 8.6 | 4.05 | 4.9 | 3 | 2.0 | 5.0 |
| $Na_2O$ | 12.4 | 14.2 | 13.19 | 8.8 | 12.5 | 11.28 | 12.5 |
| $B_2O_3$ | 18.6 | 17.8 | 13.56 | 22.0 | 14.41 | 18.47 | 14.2 |
| $Fe_2O_3$ | 6.4 | 4.0 | 1.601 | 2.8 | 2.96 | 2.96 | 2.96 |
| CaO | - | - | 4.10 | - | 2 | 3.76 | 4.1 |
| MgO | - | - | 0.348 | - | - | - | - |
| $MoO_3$ | 2.47 | 2.09 | 1.592 | 2.62 | 1.7 | 1.7 | 1.7 |
| $Li_2O$ | - | - | - | - | 2.0 | - | 2.0 |
| S.W.P.* | 15 | 12.7 | 10.9 | 15.9 | 12.22 | 12.22 | 12.22 |

* Simulated Waste Products

Table 1. Nuclear Waste Compositions (Weight %).

The solubility limited-multiple layer models imply that certain compositional ranges may exist where glasses can form a sequence of stable and continuous films capable of greatly inhibiting diffusion of other species through them or dissolution of the matrix underneath them. Results of several nuclear waste glass compositional studies show that there is a critical range of glass composition where leaching in minimal.[32-35] For example, one of the studies[33] comparing seven glasses (Table 1) for solidification of French nuclear wastes showed that glass surface attack and leach rates were strongly dependent on the concentration of network formers in the glass, Fig. 3. On the ordinate of Fig. 3 is plotted the number of days required for a glass surface to be attacked sufficiently for its IRRS spectrum at 950 $cm^{-1}$ (Si-O-alkali region) to be decreased by 15%. The abscissa is the sum of various combinations of glass network formers. The graph shows that a critical concentration of $SiO_2$ + $Al_2O_3$ + $Fe_2O_3$ > approximately 55% (by weight) or $SiO_2$ + $Al_2O_3$ > 53% must be present in the glass for the surface to be resistant to leaching. Comparison of pH changes and ion solution concentrations after 28 days, 90°C yield the same conclusion.

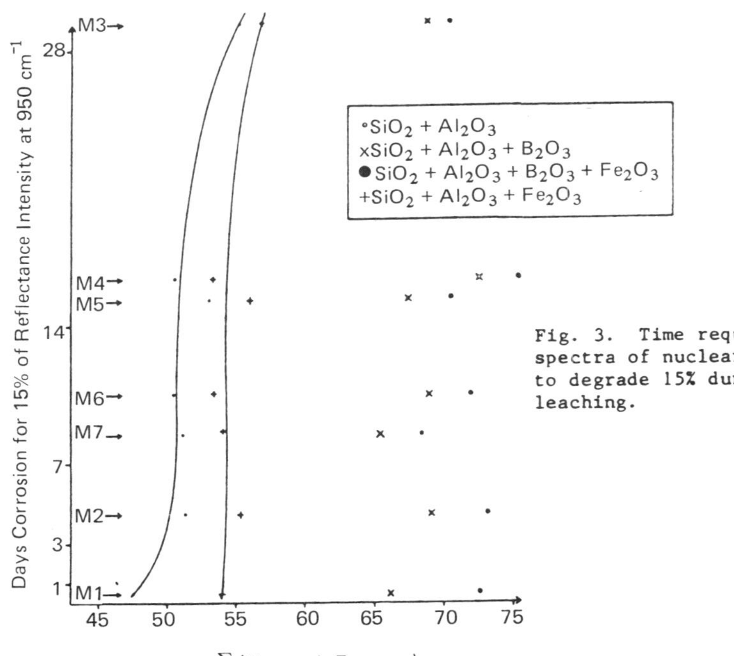


Fig. 3. Time required for IRRS spectra of nuclear waste glasses to degrade 15% during static 90°C leaching.

Another study investigated the relative importance of ZnO and $Fe_2O_3$ in stabilizing the surface film on alkali-zinc-borosilicate glasses[34]. For the composition studied, replacement of half of the 6% ZnO with $Fe_2O_3$ degraded the leach resistance by 3X. Both glasses appeared to be protected by dual films, one rich in $SiO_2$ and a second very thin film rich in multivalent species. The second film containing the mixture of ZnO and $Fe_2O_3$ was less stable than ZnO alone, suggesting critical concentrations of film formers are necessary to produce a film sufficiently dense and continuous to serve as an effective diffusion barrier.

McVay and Buckwalter have shown that the role of Fe in leaching can be especially complicated when iron silicate colloids form in the leachant.[36] Formation of iron silicate precipitates effectively removes many elements from solution and therefore eliminates many of the effects of ground water chemistry. Since the precipitates retard the saturation effects by removing species from solution, higher sustained glass leach rates and greater total elemental removal from the glass is observed. Leaching in the presence of solid pieces of ductile iron give rise to this effect and excessive concentrations of Fe in the glass may also, since high waste glasses containing large ionic Fe concentrations leaches more than 10x faster than a high Al waste.[37]

In order to generalize composition effects of Type III B glasses a computerized comparison of the compositional dependence of 27 nuclear waste glasses studied worldwide has been conducted.[35] Figure 4 illustrates the findings. All 27 glasses were leached in DI water using MCC-1 static leach procedures at 90°C for 28 days. The glass compositions were divided into 3 groups; the weight percentage of oxides of Si, B, and Na are located at the top and right corners of the ternary plot. The oxides of Al, Fe and all other constituents, labeled as WP ("waste products") are added together and comprise and 3rd axis. All of the glasses in the data file were enclosed within the compositional space labelled 45.0 in Fig. 4. Thus all nuclear waste glasses stored have a leach rate for Si at 28 days of less than 45 $g/m^2/day$. A narrower field of compositions, but still extensive, possess leach rates of 0.1 to 0.2 $g/m^2/day$. However, only a few compositions exhibit the very low range of 0.02 $g/m^2/day$ and these all contain nearly equivalent 51-53 w/o $SiO_2$ and ±2% of the other constituents.

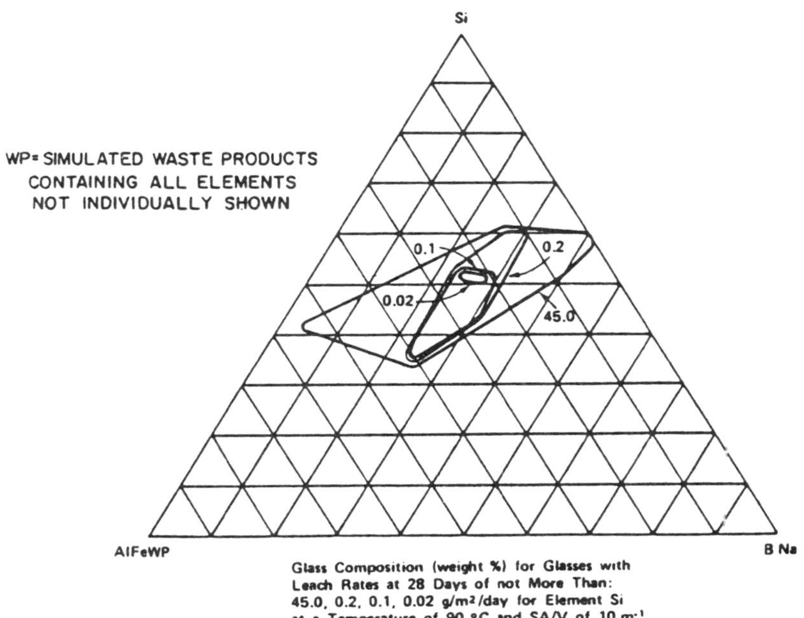


Fig. 4. Compositional dependence of nuclear waste glass leaching.

The results of Figure 4 for Si in solution are similar to the general findings of this study; i.e., there is a narrow compositional range where nuclear waste glasses achieve leach rates as low as 0.02 $g/m^2/day$ in a static MCC-1, 90°C leach test. However, a much broader range of glass compositions will yield leach rates of 0.1 $g/m^2/day$. The best compositional range tends to be approximately 51-53% ($SiO_2$), 24-28% ($Na_2O$ + $B_2O_3$), and 21-25% ($Al_2O_3$ + $Fe_2O_3$ + WP). Analysis of the time dependent changes of the compositional space for lowest leach rates leads to the conclusion that a critical concentration of $Al_2O_3$, $Fe_2O_3$, and waste product constituents are needed to produce low leach rates. This is consistent with surface analyses that show many of these species being incorporated in surface films which serve as diffusion barriers[31,36-40] slowing down long term release rates.

It has also been shown that the thickness of the Type IIIB films decreases with increasing percentage of waste constituents and multiple valence oxides.[29] Leach rates for all elements are much lower for the glasses with thinner films which indicates not only that the films formed quickly but that they were sufficiently continuous and dense to prevent congruent attack of the glass. It is especially important that burial studies of nuclear waste glasses in deep Swedish granite with 90°C centerline heaters have shown that Type IIIB surfaces form under conditions expected in long term geologic repositories.[41,42] The presence of metallic overpack interfaces such as Pb, Ti, or Cu, accelerate formation of the multiple layers apparently due to high surface area to solution volume ratios which lead to rapid precipitation of insoluble metallic compounds.[42]

Summary

Type IIIB, multiple barrier surfaces constitute an important new class of glass surface reactions. A number of alkali borosilicate nuclear waste glasses that exhibit Type IIIB surface behavior have elemental leach rates as low as 0.02 to 0.2 $g/m^2/day$ with a time dependence of static leaching of $t^{0.5}$ to $t^{0.2}$ or less after 28 days at 90°C. A discussion of the variables controlling formation of Type IIIB surfaces follows.

Although a short (several days) period of alkali-hydrogen ion exchange may occur for IIIB glasses, the dominant, long term mechanisms controlling corrosion is a combination of matrix dissolution followed by incongruent dissolution and solution/precipitation reactions. The extent of matrix dissolution and onset of surface precipitation will depend on the time required for various species in the glass to reach saturation in solution. Saturation of species (i) will be a function of the initial solution pH, amount of alkali in the glass and rate of alkali release, temperature, initial concentration of species (i) in the solution, SA/V which influences solution concentration, or flow rate which also affects solution concentration. Until saturation of some species in solution is reached, the glass dissolves congruently at a rate proportional to $kt^1$.

When solution saturation of species (i) is reached there is no longer any driving force for that species to leave the glass surface. Consequently species (i) will accumulate at the glass-solution interface as the matrix dissolves, leaving species (i) behind in the glass. If the matrix dissolution releases alkali ions, as will be the case for most glasses, there will be a concomitant rise in pH proportional to the flow rate or SA/V of the system. An increase in pH can have several simultaneous effects on the glass, the solution, and the glass-solution interface. At the new pH, a second species (j) may reach solution saturation and subsequently be retained in the glass surface along with species (i) The extent of incongruent dissolution of the glass is thereby increased. In addition, the pH can have either one of three effects on species (i), previously in saturation; 1) it remains saturated but at a higher concentration; 2) it becomes supersaturated and precipitates either on the glass or other surface or as a colloid; or 3) it becomes undersaturated and species (i) in the glass surface once again begins to be released. The sequence of events that occurs is predictable based upon the solubility limits of each species at a given pH, as shown by Grambow[28] Figure 5, based upon Grambow's work, shows that the $Fe(OH)_3$ solubility limit should be exeeded over a broad range of pH and therefore nuclear waste glasses containing Fe oxides should concentrate Fe within surface layers. Zinc, Nd, Sr, and Ca should be concentrated as well in nearly neutral or slightly alkaline solutions with Na and B depleted. Figures 6 and 7 obtained by electron microprobe analysis and AES profiling of corroded nuclear waste glasses,[29,38] shows the concentration of Fe (Fig. 6) and other heavy metal species (Fig. 7) in the glass surface. Many other studies confirm the prediction of solubility limited surface layers.[11,29-31,36-42]

A consequence of the formation of the multiple barrier, IIIB films is low overall leachability of many nuclear waste glasses over a pH range from 4.5 to 9.5. Figure 5 superimposes a plot of Si leachability from a Savannah River Lab composite waste glass immersed in a 5 day static 23°C solution buffered to various pH values from 3.5 to 10.7. Wicks[43] data shows that over the pH range expected for repository ground waters,[43] shown by arrows, glass leachability is lowest. Thus, we can conclude that the solubility limits that establish the equilibrium ionic concentrations for the ground waters should also establish the multiple barrier films to protect nuclear waste glasses in contact with those ground waters.

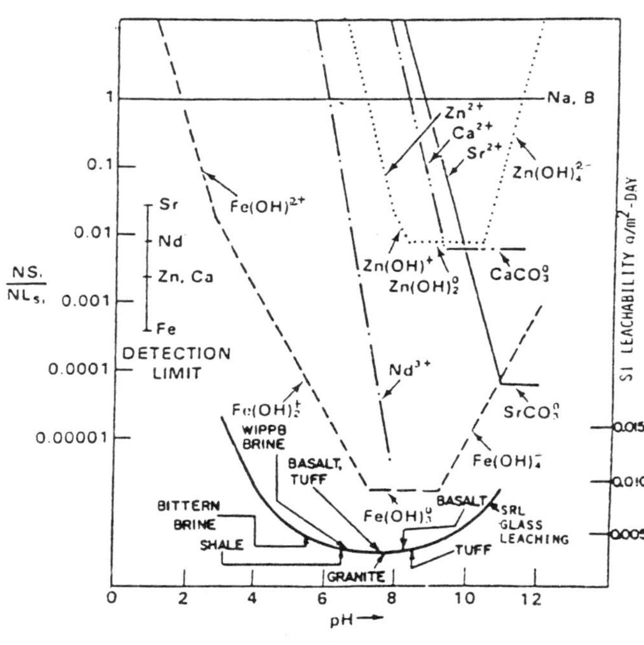


Fig. 5. Correlation between solubility limits of nuclear waste glass surface reaction products and 23°C glass leach rates in different pH solutions.

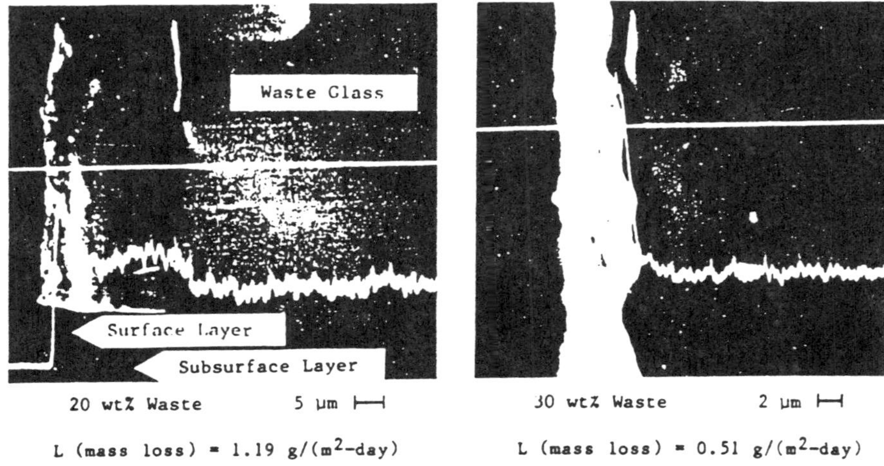


Cross sections of leached layers for 20% and 30% waste loaded glasses with x-ray line profiles of waste constituent Fe.

Fig. 6. Fe concentration in leached nuclear waste glass surface layers.

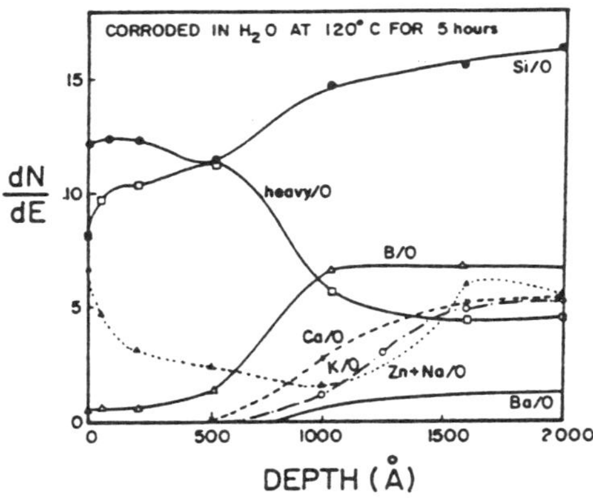


Fig. 7. Surface chemical composition profiles of a leached nuclear waste glass.

## References


1. HENCH, L. L., Survey Lecture for the XI Intl. Cong. Glass, Prague Czechoslovakia (1977).
2. HENCH, L. L., and CLARK, D. E., J. Noncrystal. Sol., 28, (1978) 83-105.
3. SCHMIDT, Yu. A. and DUBROVA, S. K., Steklo i Keram., 11, (1954) 3-7.
4. SANDERS, D. M. and HENCH, L. L., J. Amer. Ceram. Soc., 56, (1973) 373-377.
5. RYND, J. P., and RASTOGI, A. K., Amer. Ceram. Soc. Bull., 53, (1974), 631-634, 637.
6. CHAPPELL, R. A., and STODDART, C.T.H., Phys. Chem. Glasses, 15, (1974) 130-135.
7. PANTANO, Jr., C. G., DOVE, D. B., and ONODA, Jr., G. Y., J. Non-Crystal. Solids, 19, (1975), 41-53.
8. CLARK, D. E., PANTANO, Jr., C. G. and HENCH, L. L., Books for Industry, New York (1979).

9. GOSSINK, G., de GREFTE, H.A.M. and WERNER, H. W., J. Amer. Ceram. Soc., 52 (1979) 4-9.
10. LANFORD, W.A., DAVIS, K., LAMARCHE, P., LAURSEN,T., GROLEAU, R. and DOREMUS, R. H., J. Non-Crystal. Solids, 33, (1979) 249-266.
11. CHICK, L. A., McVAY, G. L., MILLINGER, G. B. and ROBERTS, F. P., Battelle Memorial Inst., PNL Report 3465, Dec. 1980.
12. BUDD, S. M., Glass Surfaces, D. E. Day, ed., North Holland Publishing Co., Amsterdam (1975) pp. 55-64.
13. BACH, H. and BAUKE, F.G.K., Phys. & Chem. of Glasses, 15, (1974) 123-129.
14. BAUCKE, F. K., J. Non-Crystal. Solids, 14, (1974) 13-31.
15. SANDERS, D. M., PERSON, W. B. and HENCH, L. L., Appl. Spectrosc., 26, (1972) 530-536.
16. SANDERS, D. M., PERSON, W. B. and HENCH, L. L., Appl. Spectrosc., 28, (1974) 247-255.
17. CLARK, D. E., ETHRIDGE, E. C., DILMORE, M.F. and HENCH, L. L., Glass Tech., 18, (1977) 121-124.
18. PANTANO, Jr., C. G., CLARK, Jr., A. E. and HENCH, L. L., J. Amer. Ceram. Soc., 57, (1974) 412-413.
19. OGINO, M., OHUICHI, F. and HENCH, L. L., J. Biomedical Maters. Res., Vol. 14, (1980) 55-64.
20. DILMORE, M. F., CLARK, D. E. and HENCH, L. L., Amer. Ceram. Soc. Bull., 57, (1978) 1040-1044.
21. DILMORE, M. F., CLARK, D. E., HENCH, L. L., Amer. Ceram. Soc. Bull., 58, (1979) 1111-1114.
22. FUJIU, T., OGINO, M., KARIYA, M. and SCHIMURA, T., private communications.
23. STANLEY, H. R., HENCH, L. L., GOING, R., BENNETT, C., CHELLEMI, S. J., KING, C., INGERSOLL, N., ETHRIDGE, E. and KREUTZIGER, K.. Oral Med., Oral Pathology, 42 No. 5, (1976) 339-356.
24. WILSON, J. and HENCH, L. L., private communications.
25. HUDSON, G. A., and BACON, F. R., Am. Ceram. Soc. Bull. 37 (1958) 185-188.
26. OKA, Y., RICKER, K. S. and TOMOZAWA, M., J. Am. Cerm. Soc., 62, (1977) 11-12, 631-632.
27. OKA, Y. and TOMOZAWA, M., J. Non. Crystal. Solids, 42, (1980) 535-544.
28. GRAMBOW, B., 5th Int. Symp. on the Sci. Basis for Radioactive Waste Management, Berlin, June 1982.
29. WICKS, G. C. and WALLACE, R. M., private communications.
30. MENDEL, J. E. et al ,Pacific Northwest Lab Report 3802, April 1981
31. PLODINEC, M. J., WICKS, G. C. and BIBLER, N.E., Report DP-1629, May 1982.
32. HENCH, L. L. and URWONGSE, L., SKBF/KBS Report, 8-82 (1980).
33. NOGUES, J. L., HENCH, L. L. and ZARZYCKI, J., 5th International Symp. on the Scientific Basis for Radioactive Waste Management, Berlin, June 1982.
34. NOGUES, J. L. and HENCH, L. L., 5th International Symp. on the Scientific Basis for Radioactive Waste Management, Berlin, June 1982.
35. HENCH, A. A. and HENCH, L. L., private communications.
36. McVAY, G. L. and BUCKWALTER, C. Q., Submitted to J. Am. Ceram. Soc., May 1982.
37. CLARK, D. E., MAURER, C. A., JURGENSEN, A. R. and URWONGSE, L., 5th International Symp. on the Scientific Basis for Radioactive Waste Management, Berlin, June 1982.
38. HENCH, L. L., CLARK, D. E. and LUE YEN-BOWER, E., Nuclear and Chemical Waste Management, 1, (1980) 59-75 .
39. HENCH, L. L., CLARK, D. E. and LUE YEN-BOWER, E., proceedings of the conference on High Level Radioactive Solid Waste Forms, Leslie A. Casey, ed., NUREG/CP-0005, published by U.S./NRC 199-235 (1979).
40. McVAY, G. L. and BUCKWALTER, C. Q., Nuclear Technologies, Vol. 51, (1980).
41. HENCH, L. L., WERME, Lars and LODDING, Alex , 5th International Symp. on the Scientific Basis for Radioactive Waste Management, Berlin, June 1982.
42. WERME, Lars, HENCH, L. L. and LODDING, Alex, 5th International Symp. on the Scientific Bais for Radioactive Waste Management, Berlin, June 1982.
43. WICKS, G. C., Proceedings of Waste Management 1981 Conference, Tucson, Arizona, Feb. 23-27, 1981

# HYDROTHERMAL CORROSION OF LITHIA DISILICATE GLASS-CERAMICS

**Yung-Kuo Chao and David E. Clark**

*Ceramics Division, Department of Materials Science and Engineering*
*University of Florida*
*Gainesville, FL 32611*

Abstract

Hydrothermal corrosion of lithia disilicate (33L) glass-ceramics having four volume fractions of crystallinity (0, 0.2, 0.6 and 0.9) was carried out at 200°C in both vapor and aqueous phases. In general, the stability of the glass-ceramic is improved with increases in the volume fraction of crystallinity. Corrosion was more extensive on the glass samples exposed to the aqueous phase than those exposed to the vapor phase. A crystalline film consisting mainly of lithia metasilicate was formed on 33L glass after 10-hr exposure of the sample to the vapor phase. This compound is thought to be an intermediate phase during devitrification of the 33L glass. The enhanced resistance of 33L glass and glass-ceramics to vapor phase attack is partially attributed to surface crystallization of the glassy phases in these samples.

## Introduction

Glass-ceramics possess unique physical and chemical properties due to their fine crystalline microstructures.[1,2] Understanding the chemical durability of these materials in a wide range of environments is essential for predicting their long-term performance. The aqueous corrosion behavior of 33L glass has been studied extensively at temperatures below 100°C.[3-6] Recently, McCracken et al. have studied the aqueous corrosion of lithia-disilicate (33L) glass-ceramics at 90°C in acidic, neutral and basic solutions.[3] In addition to the well-established ion exchange and network dissolution mechanisms of corrosion found in glasses, they recognized the importance of phase boundary and intracrystalline attacks in 33L glass-ceramics.

The 33L system was chosen for this study because the corrosion behavior of a binary system is much less complex than that of a multicomponent system. Furthermore, 33L glass-ceramics offer the advantage that their compositions are stoichiometric with respect to both the crystalline and glassy phases. 33L glass and its glass-ceramics (20%, 60% and 90% volume fractions of crystallinity) were investigated under hydrothermal conditions of 200°C and about 185 psi in both vapor and liquid phases. These conditions, which are much more severe than those normally encountered in everyday uses, are expected to result in accelerated corrosion of the samples. Increases in temperature are known to cause marked rises in the rates of reactions. Increases in pressure, likewise, accelerate corrosion, but to a lesser degree than increases in temperature.[7,8] However, it has also been observed that hydrothermal conditions promote the devitrification of amorphous silica.[7] If a glass undergoes surface crystallization under hydrothermal conditions, it could become more resistant to corrosion since crystallinity improves chemical durability.[3]

## Experimental Procedure

The $Li_2O \cdot 2SiO_2$ (33L) glass and its glass-ceramics were prepared from reagent grade $Li_2CO_3$ and 5 µm Min-u-sil. The glasses were melted in a Pt crucible inside an electric furnace maintained at 1350°C for 24 hrs. Glass patties were cast onto a graphite mold and annealed at 450°C for 6 hrs. 33L glass cylinders were nucleated at 475°C for 24 hrs and crystallized at 550°C for various times to form cylinders having 0.2, 0.6 and 0.9 volume fraction crystallization. Quantitative stereology was used to determine the percentage volume fractions of crystallization.[9] Corrosion samples were obtained by slicing the glass rods with a diamond wafering saw into pieces having approximate dimensions of 1 x 0.5 x 0.3 cm. These specimens were polished with 180-, 320-, and 600-grit SiC papers, rinsed with acetone, and dried in air.

Corrosion of the specimens was carried out at 200°C and about 185 psi in an autoclave illustrated schematically in Fig. 1. It took approximately 50 min. for the autoclave to reach the specified values of temperature and pressure. As shown in the top of Fig. 1, the sample was placed horizontally on teflon matting below the water for aqueous corrosion, and placed vertically in teflon matting above the water level for vapor phase corrosion. The bottom portion of the diagram shows the configuration inside the autoclave in which the teflon cells containing the specimens were placed on a steel support and immersed about halfway in water.

Specimens were corroded above and below deionized water in teflon cells (30 ml) placed in the autoclave for periods of 1 and 10 hrs. For comparison, 33L samples also were corroded aqueously at room temperature for 10 hrs, 33L glass at 90°C for 9 hrs, and 33N glasses (33 mol% $Na_2O$ - 67 mol% $SiO_2$) in the autoclave in both vapor and aqueous phases for 10 hrs. All runs were in deionized $H_2O$ and had a SA/V (surface area of sample to volume of solution ratio) = 0.1 $cm^{-1}$. After the cells were taken from the autoclave, the samples were rinsed, dried with acetone, and stored in a desiccator until analyzed with IRRS (infrared reflection spectroscopy), SEM and X-ray diffraction. The pH values of solutions in the cells were measured after cooling to room temperature, and the concentrations of alkali ions and $SiO_2$ were determined using atomic absorption, emission spectroscopy and colorimetry.

## Results and Discussion

Scanning electron micrographs of 33L glass and glass-ceramics corroded under autoclave conditions for 10 hrs are shown in Figs. 2-8. Fig. 2 shows the severely corroded surface of 33L glass that was immersed in D.I. $H_2O$. The porous morphology is indicative of extensive alkali leaching and corresponds to a non-protective Type IV surface as discussed by Hench and Clark.[10] A similar sample exposed only to the vapor phase exhibits a film with plate-like crystals (Fig. 3), corresponding to a Type III surface.[10] Smaller crystals are formed on the 33L-0.2 glass-ceramic exposed to the vapor phase as shown in Fig. 4. The relatively smooth upper portion of the micrograph in this figure was the area of the sample in contact with the teflon matting during corrosion. When the volume fraction of crystals was increased to 0.6, only sparsely isolated corrosion products were formed on the surface as can be seen in Fig. 5. Figs. 6-7 are micrographs of 33L-0.9 glass-ceramic corroded in aqueous and vapor phases, respectively. The surface morphology of Fig. 6 is characteristic of 33L glass-ceramic after aqueous corrosion showing both phase boundary and intracrystalline attack.[3] Fig. 7 shows the vapor phase corroded surface of the 33L-0.9 sample, which is very similar to that of a freshly abraded sample with visible polishing streaks. Thus, the SEM micrographs of the vapor phase samples (Figs. 3-5 and 7) show that the degrees of surface crystallization (resulting from hydrothermal exposure) of samples is proportional to the amount of glassy phase they contain.

Fig. 8 is the IRRS spectra of 33L specimens exposed to 200°C in the autoclave and 25°C, both for 10 hour periods. Due to surface roughness caused by extensive surface degradation, samples placed above and below water at 200°C do not give good reflectance peaks. In contrast, the sample corroded at 25°C gives rise to a large peak due to Si-O-Si bridging stretching bonds at approximately 1100 $cm^{-1}$, and a Si-O nonbridging stretching peak at approximately 900 $cm^{-1}$. The shape and position of these peaks, compared to the equivalent ones for the uncorroded samples, indicate that significant ion exchange reactions between alkali ions and $H_3O^+$ ions have occurred in this specimen. Fig. 9 is the IRRS spectrum of the 0.9 volume fraction 33L glass-ceramic corroded under the same conditions. Due to the better chemical stability of the 0.9 glass-ceramic compared with that of glass (33L), there is little change in spectrum of the sample corroded above water compared with that of the uncorroded sample. The spectra of the samples immersed in water at 25°C and 200°C show marked decreases in peak intensities compared to the uncorroded samples, again due to surface roughness.

Table I gives the experimental X-ray diffraction data for the crystalline film on the 33L glass shown in Fig. 3, and literature data (JCPDS) on lithia metasilicate $Li_2O{\cdot}SiO_2$. Based on comparison of the peak positions, the major component of the film is $Li_2O{\cdot}SiO_2$. Minor constituents could not be identified unambiguously from the remaining peak positions, but a likely candidate is $LiOH{\cdot}H_2O$.

Hench et al. have reported that a metastable lithium metasilicate appears as a precursor to equilibrium crystallization of lithium disilicate glass in air.[11] Furthermore, Sanders and Hench[6] have shown that devitrification of the 33L glass can be achieved in an autoclave at 210°C for 93 hrs, which is at least 200°C lower than required in air. Crystalline lithium disilicate was shown to be formed more rapidly than would be possible without steam. However, in the present study a crystalline film consisting mostly of $Li_2O \cdot SiO_2$ appeared on 33L glass corroded in the vapor phase at 200°C for 10 hrs based on the X-ray data in Table I. Thus, it appears that $Li_2SiO_2$ is an intermediate phase in the crystallization of 33L glass at 200°C in steam.

Table II summarizes the solution data after corrosion. The initial pH of D.I. water was about 5.7. The relatively unchanged pH values associated with samples corroded at room temperature suggests little corrosion. The tabulated solution concentrations of $SiO_2$ of these samples are probably not accurate due to the relatively large variations in measured values caused by lower sensitivity of the colorimetric method at these low $SiO_2$ concentrations. The % weight loss data of these samples are also not accurate for the same reason. On the other hand, samples exposed to the solutions having high post-corrosion pH values ( ∿11) have undergone considerable corrosion. The negative weight loss associated with some samples indicates that those samples have gained weight. The weight gain of 33L glass in vapor phase at 200°C for 10 hr is probably due to trapping or inclusion of water in the corrosion film on the glass. The 40.94% wt loss of 33L glass corroded at 200°C for 10 hrs was due to both aqueous dissolution and subsequent mechanical deterioration of the sample prior to weighing. The solution analysis, together with SEM and IRRS data of 33L glass and glass-ceramics, show that aqueous-phase attack is more severe than that in vapor phase, and that the degree of corrosion of a sample in a particular phase is proportional to time and temperature and inversely proportional to volume fraction of crystallinity (with the exception that the 0.6 glass-ceramics being more durable than the 0.9 glass-ceramics after 10h exposures). 33N samples were obviously much less durable than those of the 33L system since these samples were completely dissolved, but some powdery residues were found in the corrosion cells after cooling to room temperature.

The magnitude of the parameter, $\alpha$ (as defined by Sanders[4]), is indicative of the rate-controlling corrosion mechanism.[4,5,12] Sanders'[4] equation of $\alpha$ is in error and should be multiplied by a factor of 2. Selective leaching is operative when $\alpha$ is close to 0 and congruent dissolution is operative when it is close to 1 for glasses. Values of $\alpha > 1$ are thought to be due to either preferential adsorption [13] or precipitation during glass corrosion. For glass-ceramics, the value of may be additionally influenced by other modes of attack. Small compositional variations in the parent glass may become increasingly important as the volume fraction of crystallization increases, resulting in a phase boundary glass that is compositonally much different from the parent glass. Since $\alpha$ values are calculated based on the parent glass composition, large errors in $\alpha$ may be obtained for glass-ceramics containing large volume fractions of crystallization where phase boundary attack is usually the rate controlling mechanism.

Previous studies involving corrosion behavior of silicate glasses in both vapor and liquid phases have shown that vapor-phase attack is more severe.[14,15] Charles[14] attributed the faster hydration of a soda-lime glass in steam than that in water to the more rapid increase of pH in vapor phase. Clark and Ethridge[15] explained the vapor-phase attack of a glass enamel in a boiling 20% HCl solution in terms of the high SA/V ratio together with a continuously replenished condensate which enhanced network dissolution, and minimized solubility effects. However, for 33L glass and glass-ceramics an opposite effect (i.e., the samples are more resistant to vapor-phase attack) is observed as illustrated by the above SEM, IRRS and solution data. Comparisons of these data suggests that surface crystallization (due to hydrothermal conditions) plays an important role in the marked corrosion resistance in the vapor phase. This is illustrated by contrasting the solution data of 33L and 33N glasses at 200°C for 10 hrs. in Table II. 33N glass samples dissolved completely in both aqueous and vapor phases. 33L glass also has undergone considerable corrosion in the aqueous phase, but not in the vapor phase due to the formation of the crystalline, protective surface. If limited transport to and away from the vaor phase samples were the major reason for the reduced corrosion rates in the vapor phase, then both the 33L and 33N vapor phase samples would be expected to behave similarly. Therefore, surface crystallization must be responsible for the improvement in chemical durability.

When comparing the effects of surface crystallization in aqueous and vapor phases with respect to the volume fractions of crystallinity, quite large differences are observed for 33L samples at 200°C for 10 hrs in Table II. The effect is most pronounced for 33L glass where a large difference is observed either in $Li^+$ and $SiO_2$ concentration or % weight loss data between the two phases. However, this effect decreases rapidly as the % volume fraction of crystallinity increases as can be seen in the 33L-0.6 or 33L-0.9 glass-ceramics. Therefore, the effects of surface crystallization become less important as the volume fraction of bulk crystallization is increased.

## Conclusion

Hydrothermal corrosion of binary alkali silicate systems in both vapor and aqueous phases has been investigated. Time, temperature, and crystallinity are varied for comparisons. In vapor phase corrosion of 33L glass the formation of a crystalline film comprised mainly of $Li_2O \cdot SiO_2$ is formed and is considered as an intermediate in the devitrification process of 33L glass. Values of $\alpha$ greater than one are attributed to adsorption of ions or compositional deviations in the parent glasses. Vapor phase attack is less severe than that of aqueous phase for 33L glass and glass-ceramics under hydrothermal conditions. This behavior is reverse to that observed for glasses containing sodium ions and can be attributed to the surface crystallization of the former samples at high temperatures. The resistance to both vapor phase and aqueous phase attack generally increases as the volume fraction of crystallization increases up to 0.9. The one exception appears to be that the 0.6 glass-ceramic was more durable than the 0.9 glass-ceramic after 10h exposure.

## Acknowledgements

The authors are grateful for the support of the Air Force Office of Scientific Research under Contract #49620-80-C-0047.

## References

1. P.W. McMillian, Glass-Ceramics, Academic Press, Inc., London, 1964.

2. A.E. Berezhonoi, Glass-Ceramics and Photo-Sitalls, Plenum Press, New York, 1970.

3. W.J. McCracken, D.E. Clark and L.L. Hench, "Aqueous Durability of Lithium Disilicate Glass-Ceramics," Am. Ceram. Soc. Bull., 61 [11] 1218-23 (1982).

4. D.M. Sanders and L.L. Hench, "Mechanisms of Glass Corrosion," J. Am. Ceram Soc. 56 [7], 373-77 (1973).

5. E.C. Ethridge, Ph.D. Dissertation, University of Florida (1977).

6. Sanders & Hench, "Environmental Effects on Glass Corrosion Kinetics," Am. Ceram. Soc. Bulletin, 52[9] 662-69 (1973).

7. R.K. Iler, The Chemistry of Silica: Solubility, Polymerization, Colloid, and Surface Properties, and Biochemistry, p. 45, John Wiley and Sons, Inc., New York, 1979.

8. G.G. Wicks, W.C. Mosley, P.G. Witkop and K.A. Saturday, "Durability of Simulated Waste Glass - Effects of Pressure and Formation of Surface Layers," in Glass Microstructure Surface and Bulk, pp. 413-28, ed. by G. E. Rindone et al., North-Holland Publishing Co., Amsterdam, 1982.

9. S.W. Freiman, "Applied Steriology," in Characterization of Ceramics, pp. 555-79, by L.L. Hench and R.W. Gould, Marcel Dekker, New York, 1971.

10. L.L. Hench and D.E. Clark, "Physical Chemistry of Glass Surfaces," J. Non-Crystalline Solids, 28, 83-105 (1978).

11. L.L. Hench, S.W. Frieman and D.L. Kinser, "The Early Stages of Crystallization in a $Li_2O$-$2SiO_2$ Glass," Physics and Chemistry of Glasses 12 [2] 58-63 (1971).

12. A. Schmidt, Structure of Glass, Vol. 1., pp.253-54, translated from Russian by E.B. Uvarov. Consultants Bureau, New York, 1958.

13. R.W. Douglas and T.M.M. El-Shamy, "Reactions of Glasses with Aqueous Solutions," J. Am. Ceram. Soc., 50 [1] 1-8 (1967).

14. R.J. Charles, "Static Fatigue of Glass, I," J. Appl. Phys., 29 [11] 1549-53 (1958).

15. D.E. Clark and E.C. Ethridge, "Corrosion of Glass Enamels," Am. Ceram. Soc. Bull., 60 [6] 646-49 (1981).

Table I. Comparison of X-ray Diffraction Data on 33L Glass with the Standard Powder Diffraction File (JCPDS)

| 33L Glass Sample Vapor Phase, 200°C, 10h | | JCPDS CARD #29-829 for $Li_2SiO_3$ | |
|---|---|---|---|
| d, A° | I** | d, A° | I/Io |
| 4.595* | S | 4.696 | 82 |
| 4.297 | WW | 3.302 | 100 |
| 4.092 | W | 2.708 | 57 |
| 3.983 | WW | 2.698 | 28 |
| 3.252* | SS | 2.341 | 22 |
| 2.957 | M | 2.330 | 47 |
| 2.673* | S | 2.091 | 7 |
| 2.499 | S | 2.087 | 5 |
| 2.313* | SS | 1.7740 | 4 |
| 2.070* | W | 1.7662 | 10 |
| 1.7539* | M | 1.7636 | 13 |
| 1.6461* | M | 1.6557 | 9 |
| 1.6014 | M | 1.6521 | 8 |
| | | 1.5653 | 10 |

*Peaks that correspond closely to those on JCPDS card.
**SS = very strong, S = strong, M = medium, W = weak, and WW = very weak.

Table II. Solution Data for Corrosion of 33L Glass-Ceramics and 33N Glass

| Composition | Volume % Crystal-linity | Type of Corrosion | SA/V ($cm^{-1}$) | Temper-ature (°C) | Time (hrs) | pH | $[Li^+]$ (PPM) | $[Na^+]$ (PPM) | $[SiO_2]$ (PPM) | % wt loss | α |
|---|---|---|---|---|---|---|---|---|---|---|---|
| 33L | 0 | aqueous | 0.1 | 25 | 10 | 7.50 | 1.2 | - | 1.8 | 0.10 | 0.17 |
| 33L | 20 | " | " | " | " | 7.18 | 1.2 | - | 2.0 | 0.00 | 0.19 |
| 33L | 60 | " | " | " | " | 7.05 | 0.9 | - | 3.3 | 0.05 | 0.44 |
| 33L | 90 | " | " | " | " | 7.21 | 1.0 | - | 4.0 | 0.04 | 0.47 |
| 33L | 0 | " | 0.1 | 90 | 9 | 11.57 | 24.5 | | 12.6 | 0.37 | 0.06 |
| 33L | 0 | aqueous | 0.1 | 200 | 1 | 11.65 | 108.0 | - | 585.0 | 2.03 | 0.62 |
| 33L | 20 | " | " | " | " | 11.23 | 83.0 | - | 583.0 | 2.41 | 0.80 |
| 33L | 60 | " | " | " | " | 10.44 | 33.0 | - | 318.0 | 1.26 | 1.11 |
| 33L | 90 | " | " | " | " | 10.34 | 32.0 | - | 308.0 | 1.06 | 1.09 |
| 33L | 0 | aqueous | 0.1 | 200 | 10 | 11.58 | 425.0 | - | 3250.0 | 40.94 | 0.87 |
| 33L | 20 | " | " | " | " | 11.44 | 360.0 | - | 2925.0 | 19.06 | 0.92 |
| 33L | 60 | " | " | " | " | 10.37 | 50.0 | - | 725.0 | 4.16 | 1.65 |
| 33L | 90 | " | " | " | " | 10.67 | 75.0 | - | 900.0 | 5.56 | 1.37 |
| 33L | 0 | vapor | 0.1 | 200 | 10 | 7.66 | 3.7 | - | 19.0 | -3.45 | 0.59 |
| 33L | 20 | " | " | " | " | 7.75 | 2.2 | - | 12.5 | -1.79 | 0.65 |
| 33L | 60 | " | " | " | " | 6.81 | 1.0 | - | 8.8 | 0.04 | 1.00 |
| 33L | 90 | " | " | " | " | 7.28 | 1.9 | - | 17.5 | 0.00 | 1.05 |
| 33N | 0 | aqueous | 0.1 | 200 | 10 | 11.52 | - | 3600 | 12500.0 | 100.00* | 1.31 |
| 33N | 0 | vapor | " | " | " | 11.53 | - | 3400 | 11500.0 | 100.00* | 1.28 |

*some residue remaining in corrosion cell

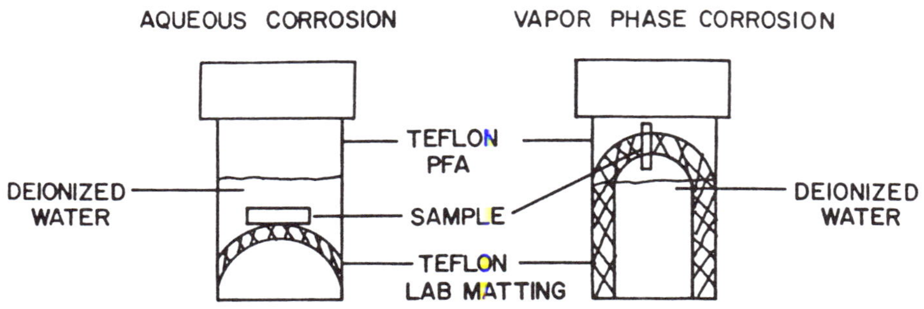


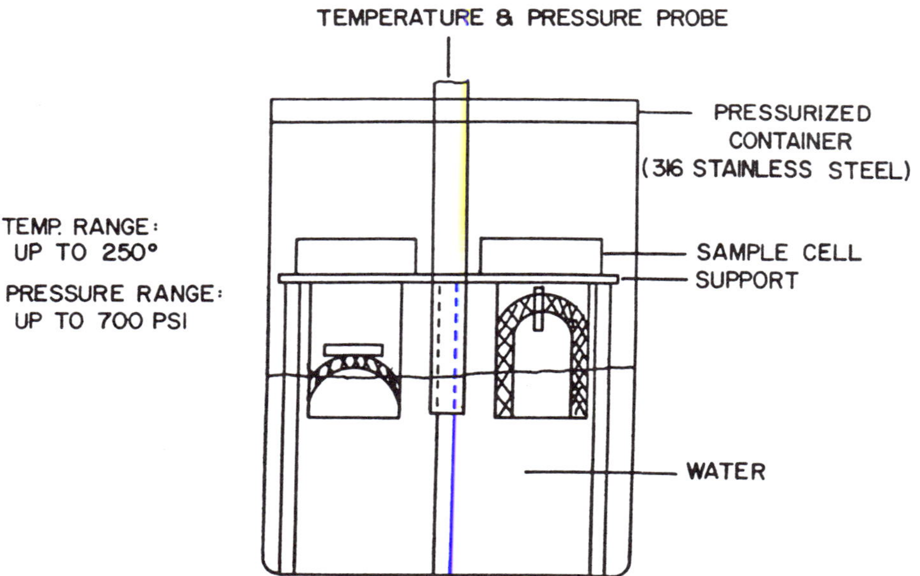


Fig. 1. Corrosion cell configuration.

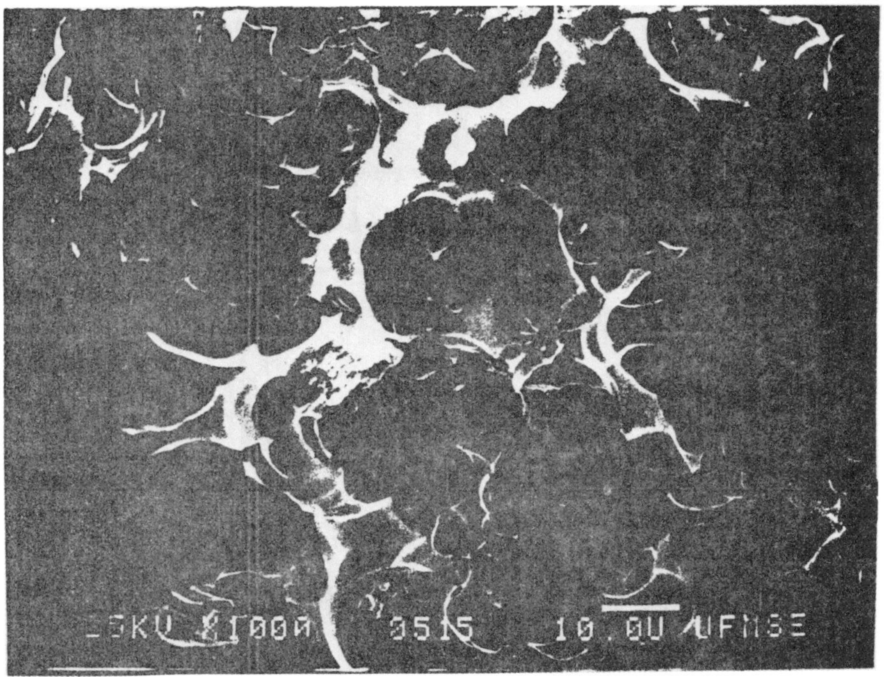


Fig. 2. Scanning electron micrograph of 33L glass after 10-h expcsure in D.I. $H_2O$ at 200°C with SA/V=0.1 $cm^{-1}$ (X1000).

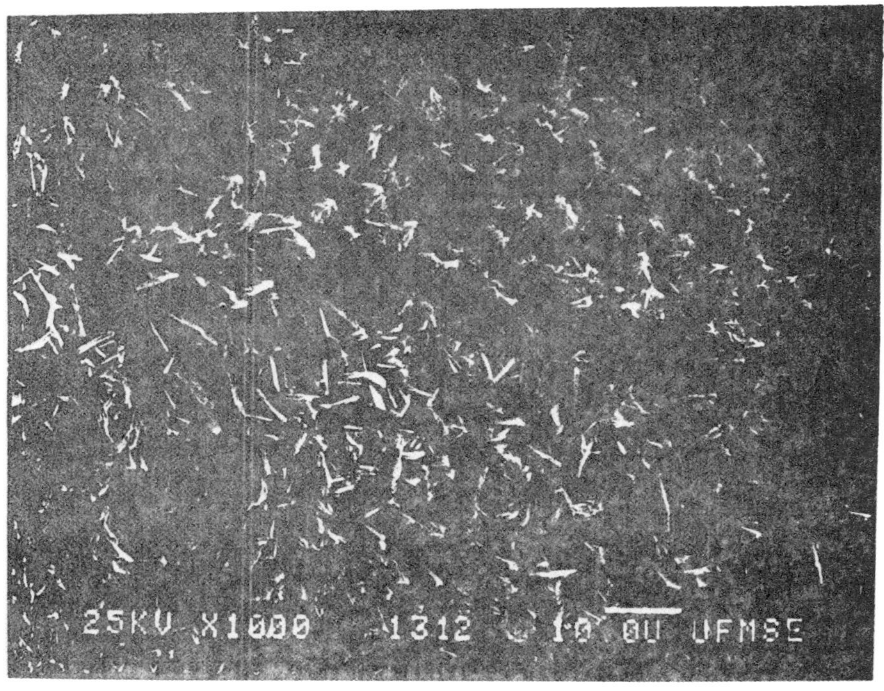


Fig. 3. Scanning electron micrograph of 33L glass after 10-h exposure above D.I. $H_2O$ at 200°C with SA/V=0.1 $cm^{-1}$ (X1000).

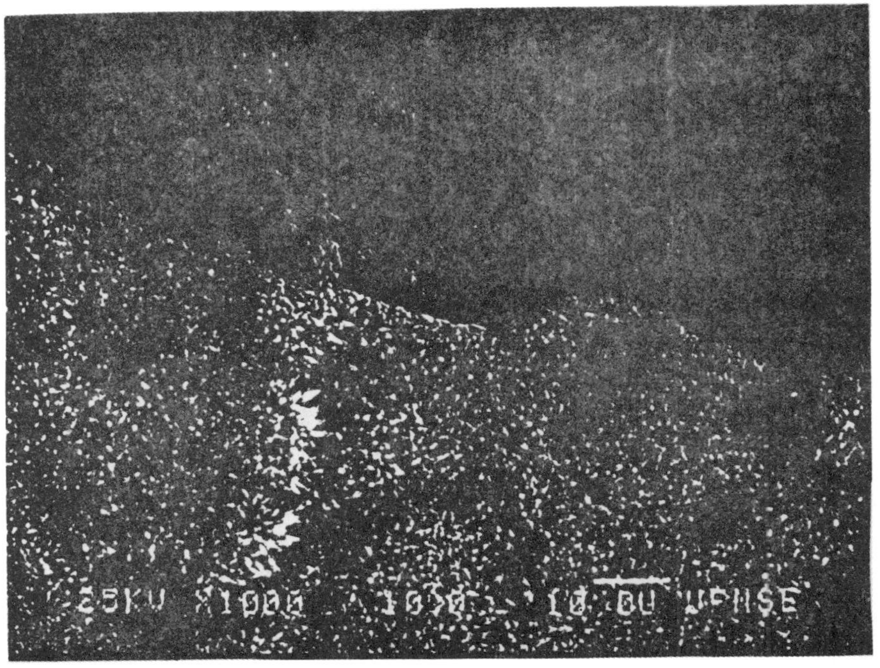


Fig. 4. Scanning electron micrograph of 33L-20% crystalline glass-ceramic after 10-h exposure above D.I. $H_2O$ at 200°C with SA/V=0.1cm$^{-1}$ (X1000).

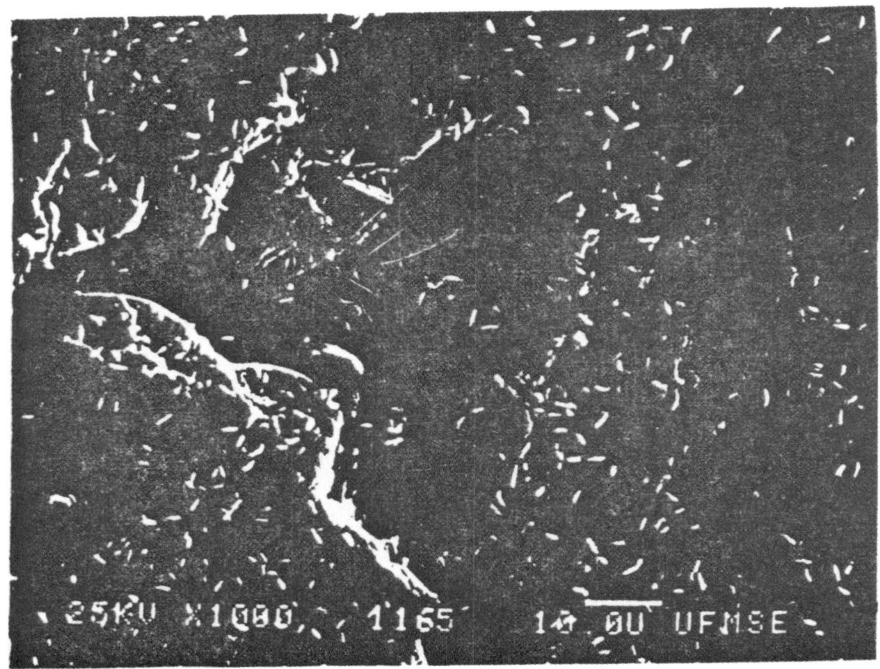


Fig. 5. Scanning electron micrograph of 33L-60% crystalline glass-ceramic after 10-h exposure above D.I. $H_2O$ at 200°C with SA/V=0.1 $cm^{-1}$ (X1000).

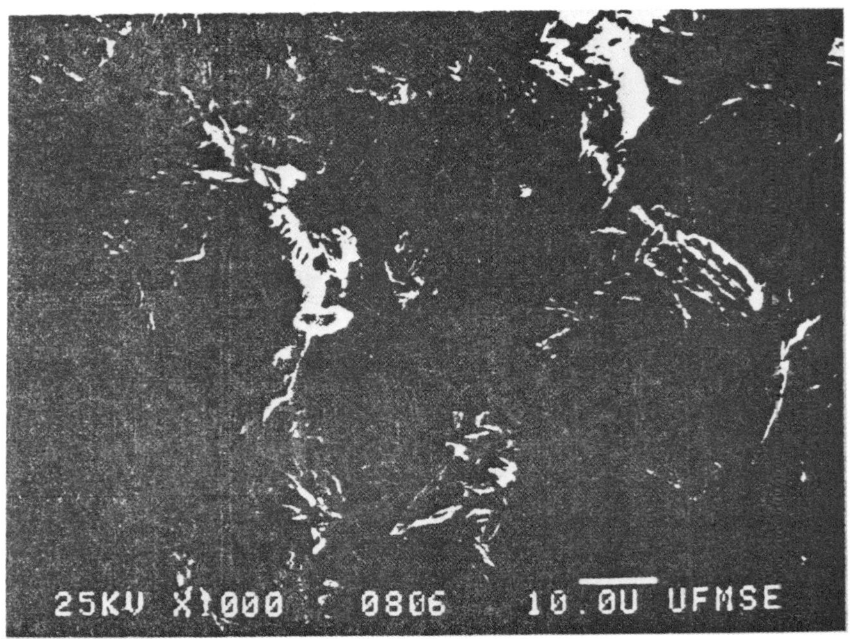


Fig. 6. Scanning electron micrograph of 33L-90% crystalline glass-ceramic after 10-h exposure in D.I. $H_2O$ at 200°C with SA/V=0.1 $cm^{-1}$ (X1000).

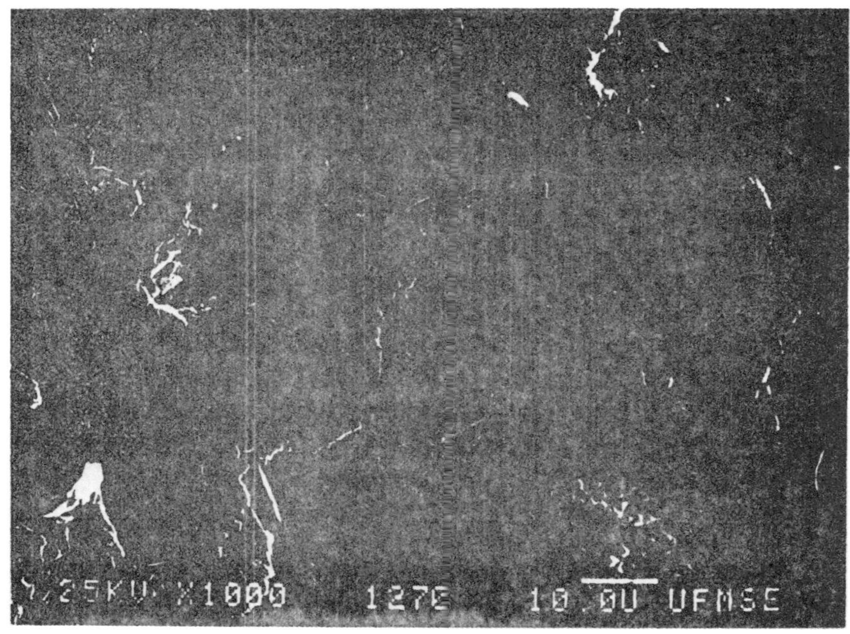


Fig. 7. Scanning electron micrograph of 33L-90% crystalline glass-ceramic after 10-h exposure above D.I. $H_2O$ at 200°C with SA/V=0.1cm$^{-1}$ (X1000).

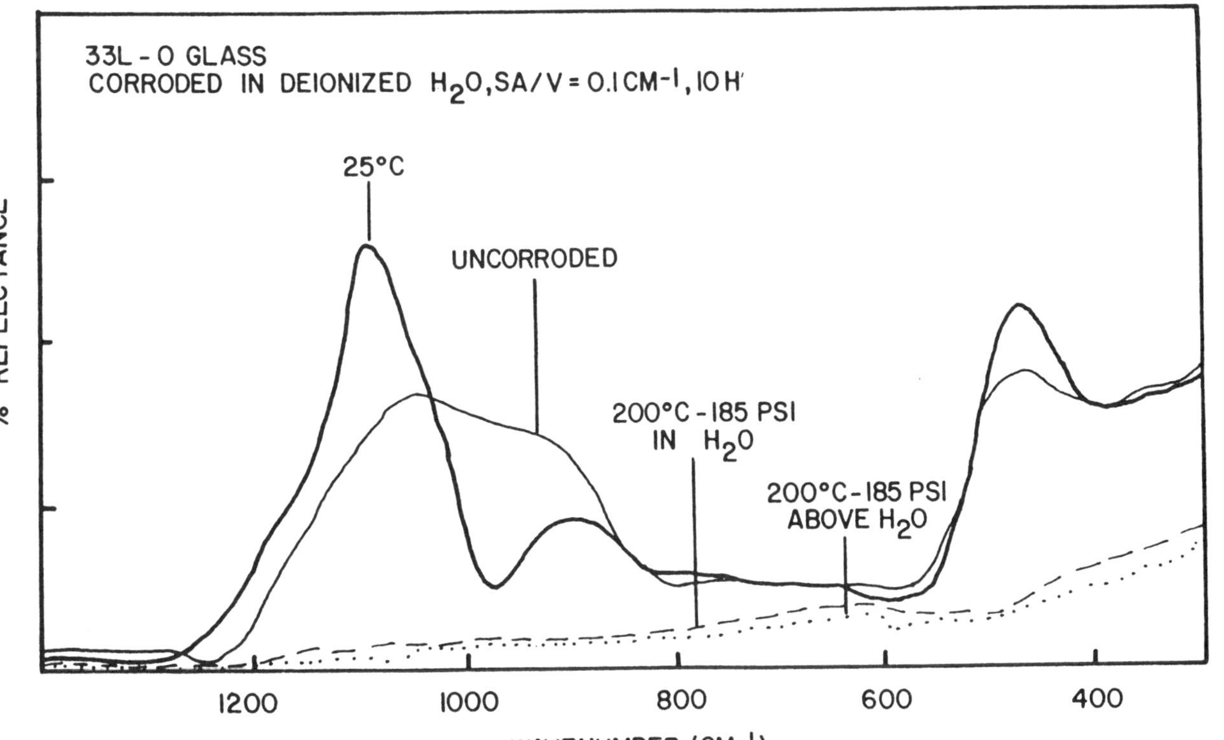


Fig. 8. Infrared reflection spectra of 33L-0 glass corroded for 10h, SA/V = 0.1 $cm^{-1}$ at 25°C and 200°C in D.I. $H_2O$ and at 200°C above D.I. $H_2O$

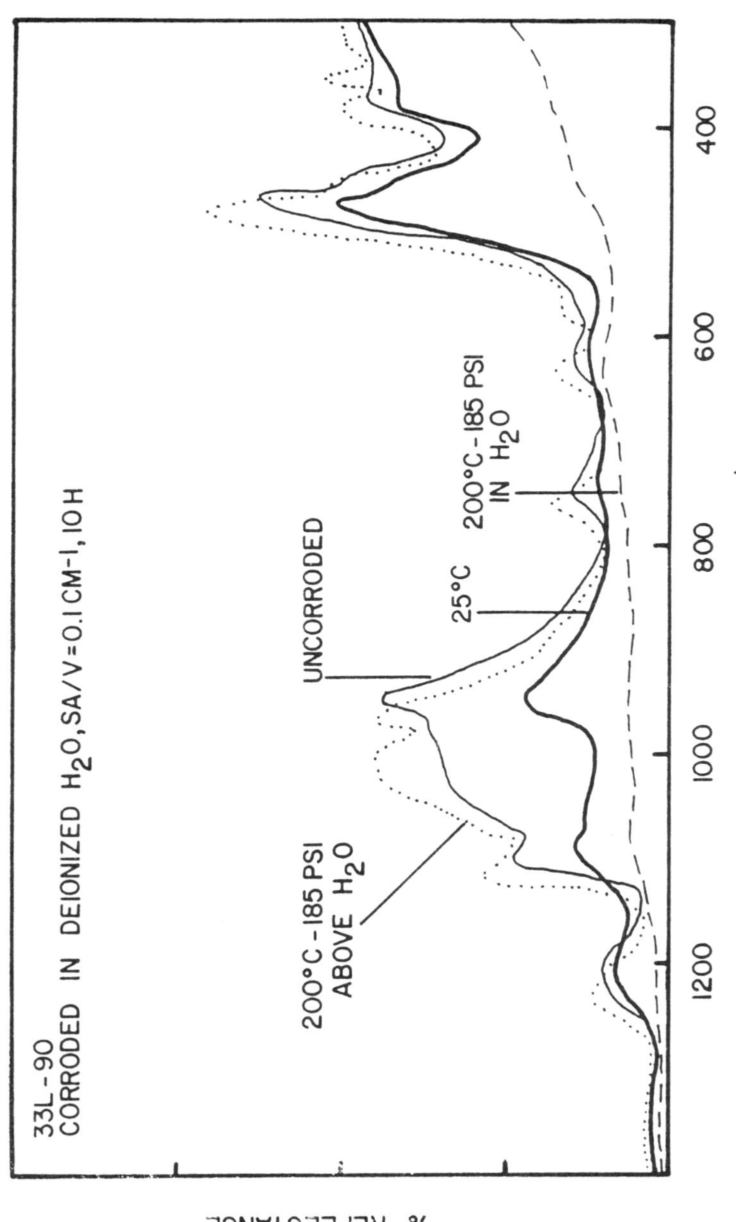


Fig. 9. Infrared reflection spectra of 33L-90 glass-ceramic corroded for 10h, SA/V - 0.1 cm$^{-1}$ at 25°C and 200 °C in D.I. $H_2O$ and at 200°C above D.I. $H_2O$.

# FRACTURE MECHANICS AND FAILURE PREDICTIONS FOR $Li_2O{\cdot}2SiO_2$ GLASS AND GLASS-CERAMICS

R.A. Palmer and L.L. Hench

ABSTRACT

Fracture mechanics techniques are used to determine the fatigue parameters for the lithium disilicate glass/glass-ceramic system. Glass-ceramics are ideal materials for examining the effects of crystallization on fatigue behavior. Two methods (dynamic fatigue and slow crack growth) are employed and are found to produce different values for the fatigue parameters. These parameters are used to design lifetime prediction diagrams and, theoretically, should be independent of the method of determination.

The limitations of the present fracture mechanics theory are discussed in the light of the inconsistencies found between these methods. It is recommended that a combination of techniques be used when reliable lifetime predictions are required. A matrix of tests, including static and dynamic fatigue and proof tests, is proposed as an improved method of characterizing the fatigue behavior of a brittle material.

## I. INTRODUCTION

For over ten years now, semi-empirical equations derived from linear elastic fracture mechanics have been employed in devising proof tests to ensure a minimum service life for components made of brittle materials. The critical parameters in these equations are determined by experiment and are assumed to be constant. This study uses two techniques (dynamic fatigue and slow crack growth) and aging in water to show the variability possible in these constants.

Subcritical crack growth in a brittle material has been shown to follow the exponential equation[1]

$$V = A\ K^{N} \tag{1}$$

where V is the crack velocity, K is the stress intensity factor (for Mode I failure in this case), and A and N are experimentally determined constants. The parameter N, termed the stress corrosion susceptibility, will be the main object of scrutiny in this work.

Derived from Eq. (1) is the equation for dynamic fatigue analysis,

$$\sigma^{N+1} = B\ (N+1)\ \sigma_c^{N-2}\ \dot{\sigma} \tag{2}$$

where $\sigma$ is the failure stress at stressing rate $\dot{\sigma}$, $\sigma_c$ is the strength of the material in an inert atmosphere, and

$$B = \frac{2}{AY^2(N-2)K_c^{N-2}}$$

where Y is a geometric constant and $K_c$ is the critical stress intensity factor.

These equations have been discussed frequently[1-5] in the literature and will not be enlarged upon here. The slopes of the logarithmic plots of Eqs. (1) and (2) yield values of N.

Having experimentally determined the parameters N and B, a proof test design diagram may be constructed by the formula

$$t_{min}\,\sigma_a^2 = B(\sigma_p/\sigma_a)^N, \tag{3}$$

where $t_{min}$ is the minimum time to failure under the applied stress $\sigma_a$ and $\sigma_p$ is the proof test stress. Using this diagram, a proof test can be designed for specific cases of applied stress and expected lifetime.

## II. EXPERIMENTAL PROCEDURE

(1) Sample Preparation.

The parent glass was made by mixing reagent grade $Li_2CO_3$ and 5μm silica sand for one hour on a roller mill in a plastic jar. Each batch weighed 200-250g and was melted for 24 hours in a covered platinum crucible in an electric muffle furnace at 1350°C.

Casting was done in a graphite mold or graphite forms. Discs required for the biaxial flexure test were made as follows: Cylinders 25mm in diameter and about 50mm long were formed and subsequently cut into 2mm slices with a diamond saw. For double cantilever beam specimens, bricks 20mm x 25mm x 80mm were poured and subsequently cut into plates with a diamond saw.

The completely glassy specimens were annealed immediately after casting for four hours at 350°C and allowed to furnace cool. Qualitative analysis of residual stresses was made using a polariscope with a tint plate. Samples showing excessive stress were remelted and recast.

Those samples to be crystallized were placed in a tube furnace directly after casting and held at the nucleation temperature. For all levels of crystallinity, the nucleation treatment was 24 hours at 475°C. This treatment was selected on the basis of the work of Freiman.[6] The size of the samples was also a factor. Because they were so large, a long nucleation time was desired to assure a consistent microstructure throughout the specimen as well as from specimen to specimen. Two cylinders or one brick could be treated at a time. To minimize thermal gradients, the specimens were held in steel wire mesh boats and set on an aluminum block in the center of the furnace.

To crystallize the specimens, the furnace was turned up to 550°C and left for various periods of time. The furnace reached the upper temperature in 15-20 minutes. Two hours resulted in about 10% crystallinity; four hours gave about 60%; fully crystallized specimens were obtained by leaving them at 550°C for 24 hours. However, the fully crystalline material was replete with microcracks which resulted in about 8% porosity (quantitative microscopy techniques[7] were used to measure the percent crystallinity).

After crystallization, the specimens were removed from the tube furnace and placed in a small furnace at 200°C to prevent thermal shock and allowed the furnace to cool to room temperature.

Discs for biaxial flexure were cut from the cylinders with a high speed diamond saw. Water was used as a coolant. Because of the rough finish and the reactivity of the materials with water, each disc was polished dry with SiC paper to a 600 grit finish. The final discs were about 2.5mm thick and were kept in a desiccator to prevent atmospheric water attack before testing.

Plates for double cantilever beam (DCB) testing were cut from the bricks to the approximate dimensions 1mm x 12mm x 75mm. These were cut using a low speed diamond wafering saw, with mineral oil or similar fluid as a coolant. Although the surface finish is not important in this test, care was taken to avoid contact with water. A groove roughly half the thickness of the specimen was machined down the center. This gives the propagating crack an easy path to follow. The groove was made using a milling machine with a SiC wheel.

(2) Test Techniques.

In Figure 1 is shown the test jig for the biaxial flexure test. The sample cup, which is mounted on the load cell of an Instron* testing machine, has three press-fitted ball bearings which define a support circle for the disc samples and also allows liquids or gases to be added for testing in various reactive or non-reactive environments. The loading pin is mounted on the crosshead of the testing machine and is centered with respect to the circle defined by the ball bearings. This method has been proposed as a standard ASTM test and evaluated by Wachtman et al.[8]

Samples were tested in biaxial flexure at five crosshead speeds over three decades (0.2, 0.1, 0.02, 0.01, and 0.002 inches per minute which correspond to 158, 79, 16, 7.9, and 1.6 $MPa\text{-}s^{1}$ stressing rates) in three atmospheres (air, water, and liquid nitrogen). Some samples were also tested after aging one day or one week in water.

Testing in air was straightforward (22°C, 65-70% relative humidity). Care was taken when testing in water to ensure that the samples tested at the fastest rate were in water for about the same length of time as those tested at the slowest rate (about two minutes).

In the aging experiments, samples were placed in plastic vials with sufficient water to achieve a surface area to volume ratio (SA/V) of 1 $cm^{-1}$. Discs were then tested in the same water in which they were aged.

Samples tested in liquid nitrogen were pre-cooled to avoid thermal shock failure. The test jig was cooled with liquid nitrogen and filled with the liquid during testing. Because of the absence of fatigue caused by aqueous attack at liquid nitrogen temperatures, testing was done only

---

*Instron Corp., Canton, MA

at one rate of 0.02 inches per minute. The average strength from this testing was used as the $\sigma_c$ value in Eq. (2).

Freiman et al.[9] give a detailed analysis of the constant moment double cantilever beam technique. The test apparatus is shown in Fig. 2. The test specimen is cemented using epoxy to the loading arms in pair of slotted inserts to ensure proper alignment. All the pivot points have suitably low friction provided by bearings. The load is applied by means of a weight pan connected through a triangular piece (assuring equal load distribution) to the loading arms. A constant load provides a constant moment applied by the arms to the specimen. This yields a constant stress intensity factor defined by

$$K = \frac{TL}{\sqrt{It}} \tag{4}$$

where T is the applied load, L is the length of the moment arm, $I = \frac{1}{12} bh^3$ which is the moment of inertia of the beam, and t is the web thickness. These terms are further described in Fig. 3.

A starter crack is initiated at the base of the slot in the groove by tightening a sharp screw against the ungrooved side. The crack will grow at a constant velocity under a constant load. The range of velocities measured was from $10^{-10}$ to $10^{-4}$ m-s$^{-1}$. With proper care, multiple measurements may be made on one sample by changing the load after taking sufficient readings at one load. Measurements were made every few hours for very slow velocities and every half minute or so for very fast velocities. Again, testing was done in air and water at room temperature (~22°C).

Measurement of the crack velocity was made with a traveling microscope.*

*Gaertner Scientific Corp., Chicago. IL

The magnification was 32X. The accuracy of the microscope was ±0.0005mm. Readings were made to the nearest micrometer.

The above procedure will provide crack velocity data for stress intensity factors less than $K_c$. Values for the critical stress intensity factor, $K_c$, were determined using this apparatus attached to an Instron** machine. An initial load was applied to begin the crack propagation, then the crosshead was turned on at a constant speed of 0.02 inches per minute and the crack was then propagated to failure. The highest value of the load (T) then is substituted into Eq. (3) and the value for $K_c$ determined.

## III. RESULTS

(1) Dynamic Fatigue.

In Fig. 4 are shown the plots of strength versus stressing rate for all materials and environments. Values of N and ln B derived from these plots and Eq. (2) are presented in Table I. The N values are found to vary with environment, crystallinity, and aging time in water. The strength (at a given stressing rate and as tested in liquid nitrogen) also changed as a function of these variables, as shown in Table II.

(2) Slow Crack Growth.

In Fig. 5 are examples of crack velocity versus stress intensity plots for 33L Glass and glass-ceramics. The fracture parameters (N and ln A) derived from these plots and Eq. (1) are given in Table III. The values for N determined in these experiments remained relatively constant compared to the values derived from dynamic fatigue data.

Values for $K_c$ were determined for 33L-Glass and 33L-92% crystalline materials only, and were found to be 0.78 MPa-$m^{\frac{1}{2}}$ and 4.19 MPa-$m^{\frac{1}{2}}$ respectively.

---

**Instron Corp., Canton, MA

(3) Proof Test Diagrams.

Design diagrams from Eq. (3) are given in Figs. 6 (dynamic fatigue data) and 7 (slow crack growth data). Because of the large variability in N, B, and A, it is very difficult to determine which plot is valid for a particular material.

## IV. DISCUSSION

(1) The Effect of Crystallization on Strength.

Reported in Table II are the average inert strength and strength values for each condition tested at 16 $MPa\text{-}s^{-1}$ (these values will serve to show the general trend). Only 33L-92% material loses strength dramatically after aging one week. Strength decreases too, for 33L-60%, but not so precipitously. This is likely to be due to phase boundary attack as described by McCracken et al.[10]

It is difficult to interpret the effect of crystallization on the strength of this system because of the microcrack formation at high Vv and the residual stress at low Vv. Previous investigators[6,11,12] have analyzed their materials' strength as a function of mean free path between dispersed particles. That is, the flaw size is limited by the interparticle spacing. Without sufficient nucleation, the crystal growth can result in formation of microcracks which greatly reduce strength.[11]

For 33L-10% crystalline, the residual stresses in the glassy phase induced by the thermal expansion mismatch between the glass and spherulites is most likely responsible for the strength increase over 33L-Glass. Freiman[6,11] found an increase in Young's modulus with crystallinity, but not enough to explain the strength increase (recall that the failure strength is proportional to $E^{\frac{1}{2}}$). Miyata and Jinno[11] re-analyzed some of the work of Hasselman and Fulrath[13] and observed a decrease in strength at low Vv (< 20%) with no connection to the mean free path. Rao[14] studying lithium disilicate containing 2-5μm spherulites, attributed the strength increase to an increase in fracture surface energy due to pinning by the crystals.

The decrease then increase in strength which occurs on exposure to water in 33L-60% is analogous to that which occurs in 33L-Glass. The residual stress merely postpones the effect.

For 33L-60% and 33L-92% crystalline, microcracks are prevalent throughout the material. It has been shown[15] that microcracking can be effective in toughening glass-ceramics. The microcracks provide multiple branches (which acting by themselves will reduce the energy of a propagating crack) leading to the spherulites which pin the crack.

The weakening on aging is due to the migration of water into the grain boundaries (via the microcracks), thus dissolving the residual glassy phase and destroying the cohesiveness of the network. Even without stress, such an interfacial attack of the glassy matrix in the spherulite crystals is observed.[10]

(2) The Effect of Crystallization on N.

The N values for all test conditions are shown in Table I. Except for 33L-Glass, the N value decreases or stays the same as the availability of water increases. This underscores the importance of the kinetics of the reaction at the crack tip.

While the glass shows an initial decrease in N from testing in air to testing in water, the subsequent increase after aging one day is indicative of another mechanism acting during aqueous corrosion.

Previous work has shown five types of glass-surface conditions, depending on the composition of both the glass and the corrosive medium.[16] One type involves the development of a protective silica-rich layer on the surface. Sanders and Hench[17] have shown that 33L-Glass undergoes this type of corrosion. Apparently this reaction layer can slow down the ion exchange process at the crack tip necessary for stress corrosion, thus resulting in a higher N value. The partially crystalline materials do not exhibit this behavior because most of the corrosive attack is along grain boundaries.[10] This intergranular attack can only be detrimental to the strength, since the "crack tip" cannot be rounded, only extended.

For the crystallized materials, N decreases or stays the same with aging in water down to the range of values for glass. This implies that (eventually) the residual glassy phase is the controlling material for stress corrosion in the glass-ceramics.

For 33L-92% crystalline, the microcracks and extensive grain boundary surface area allow rapid takeover of the process by the residual glassy phase. The most significant point here is the loss in strength after aging one week which was discussed earlier.

For 33L-10% crystalline, residual stresses in the glassy phase obviously affect the kinetics of the aqueous corrosion (the effect of this stress on the strength was discussed earlier). However, once sufficient aging has allowed penetration of glass/crystal boundaries, some of the stress is then relieved, at which point the glass corrosion again becomes the controlling factor.

For 33L-60% crystalline, although testing was not done in air or in water without aging, it appears that the same phenomena are occuring as in 33L-10% crystalline. The interaction may proceed more slowly (N = 59.2 vs. N = 46.3 after aging one day), but the presence of considerably more grain boundary area drops the N value to a lower one after one week (23.6 vs. 27.4). Microcracking is also likely to play a role in the stress corrosion process.

(3) The Effect of Crystallization on Lifetime Predictions.

Although in two cases there appear to be strength increases after aging, the proof stress test ratio ($\sigma_p/\sigma_a$) indicates there is no acvantage gained over the long term (these are reported in Table IV). Except for 33L-Glass (in water versus aged), this ratio increases with increasing availability of water. This means that to guarantee a particular lifetime, a component must be tested at a much higher proof test stress. For example, for 33L-Glass, even though testing in air and after aging one day result in about the same strength, the proof test to guarantee ten years at 50 MPa would be 140 MPa in air versus 275 MPa after aging.

Recalling Eq. (3), as N decreases, the time to failure under a given load after a given proof test decreases dramatically. Correspondingly, the proof test ratio for a given lifetime/load condition must increase. The same is true as B decreases (as in B becomes more negative). This means that even if a material becomes stronger (in short term testing) after aging, it will be the fatigue parameters, N and B, which will control the material's long term behavior.

(Note: At this point in time, the parameter B has not been well characterized. In Eq. (3), it must have the units of $MPa^2$-s, but is also has been defined in terms of $K_c$ raised to some power which depends on N. The theoretical approach employed here cannot resolve this problem.)

For 33L-Glass, the proof test stress ratio increased when tested in water versus air because of the sharp drop in N. After aging one day, the increase is due to the change in ln B.

For the crystallized materials, it is always a large change in N which is responsible for the increased ratio. For these materials, in B is almost always increasing. It may be stated ther, that for the lithium disilicate glass/glass-ceramic system, the N value controls the predicted lifetime.

(4) Slow Crack Growth.

In the slow crack growth experiments, the value of N, the stress corrosion susceptibility, was found to be unaffected by aging in water and only slightly increased by crystallization. These observations are consistent with previous investigations.

Crack front interaction with the microstructure made determination of the V-K relationship difficult for all materials except 33L-Glass. Pletka and Wiederhorn[18] also notice this effect in a magnesium alumino-silicate glass-ceramic. The crack arrest at the particle results in the stopping of the crack followed by a release with sufficient energy built up to allow the crack velocity to increase by an inordinate amount. Theocaris and Milios[19] have studied cracks propagating through a bi-material interface and have quantified the increase in velocity as a function of relative ductilities of the two materials. However, applications of their findings to materials with a brittle matrix and brittle inclusions is not straightforward.

Ferber and Brown,[20] using the double torsion method, conclude that because of crack/microstructure interactions, reliable V-K data could not be obtained on their alumina samples. Studies by Fuller[21] and Pletka, Fuller, and Koepke[22] also cast doubt on use of the double torsion method. The double cantilever beam constant moment technique is similar to the souble torsion method in that is uses a constant K specimen. However, in the DCB method employed herein, the crack growth is viewed and measured directly, which allows the microstructural effects to be monitored more closely.

By marking the locations on the sample where the arrest-release events occur, subsequent microscopic inspection of the crack surface can lead to a better understanding and perhaps quantification of the crack/microstructure interaction.

(5) Comparing Dynamic Fatigue and Slow Crack Growth.

The N values obtained in the crack growth studies correspond best to the lowest values obtained in dynamic fatigue. Pletka and Wiederhorn,[18] contrarily, found larger values for N in crack growth studies than in dynamic fatigue. Their explanation is that in a uniaxial stress field the obstacles more readily arrest the crack, yielding a higher N value in crack velocity experiments than in a four-point bend test or biaxial flesure.

For the case of the lithium disilicate system studied here, it is likely that the kinetics of the corrosion reaction at the crack tip govern the fatigue behavior rather than the macroscopic stress state. The time required for slow crack growth allows more of a reaction at the crack tip, resulting in smaller N values for the test which takes longer to complete. The short fracture time (~ ½ - 2 seconds) in dynamic fatigue is too rapid to allow sufficient degradation of the material at the microscopic flaw at least relative to the macroscopic crack. The occurence of aging effects in dynamic fatigue but not in slow crack growth confirms this interpretation.

The results of this study show that lifetime prediction diagrams drawn using crack growth data would be more conservative than those drawn using dynamic fatigue data. At this time it is not known which would be more or less accurate. The strengthening effect of water on the 33L-Glass (indeed, on most glasses) complicates the decision. It is evident that each material will need extensive testing of various types (those used here as well as static fatigue) before an intelligent decision can be made regarding lifetimes in critical applications.

In their work, Pletka and Wiederhorn[18] studied a lithium alumino-silicate glass-ceramic and found very little difference between the parameters determined by crack growth and those determined by dynamic fatigue. They attribute this to a more homogeneous microstructure compared with their magnesium aluminosilicate. This underscores the need for a variety of tests of whatever material is under investigation.

In order to use the present lifetime prediction theories, it must be remembered that the empirical model developed to describe the fatigue behavior of a ceramic under a static load assumes that the parameters derived from short term tests remain constant over the lifetime of the material. Unfortunately, some materials will react with their environment over time in such a way that the parameters originally used must change. These changes will then negate the original analysis. Each material will need to be characterized extensively as to how the parameters might change if the model is to be valid. For example, if N increases (as with 33L-Glass after aging in water one day), the original lifetime prediction will be conservative. If N decreases (as with 33L-92% crystalline after aging one week), the component's lifetime may be drastically (and catastrophically) shortened.

Thus, the present model is not designed to handle a dynamic system. A new model needs to be developed to account for changes in the parameters which describe the process of subcritical crack growth. The results of this study suggest that the following innovations should be considered:

1) the kinetics of the chemical reaction at the crack tip need to be accounted for. These studies would establish the time dependence of the fatigue parameters and (possibly) the tip radius and depth of the flaw.

2) the stress dependence of the chemical reaction at the crack tip must be established. This also affects the fatigue parameters and flaw characteristics.

3) the chemical changes in the surface layer surrounding the flaws must be characterized. A change in the chemistry (e.g. $H^+$ replacing Li or $Na^+$) or state (e.g. glass to gel) will certainly affect the growth of the flaw.

4) the synergism of all three of the above needs to be considered as well.

## V. CONCLUSIONS

1. The testing of a material which readily reacts with an aqueous environment has shown that the present model for describing fatigue by subcritical crack growth needs to be revised to account for the dynamics of interaction if the material with its environment. Because the fatigue parameters (N and B) involved in calculating lifetimes have been observed to change with time, a single analysis based on a short term test is not adequate for reliable predictions. Two further developments are needed: First, the model must be extended to account for the changes in the fatigue parameters; second, a series of tests needs to be devised to characterize these changes for a given material.

2. The strength of lithium disilicate materials increases as the percent of crystallinity increases, but not in a simple straightforward manner. At low levels of crystallinity, a compressive stress may be induced in the glassy matrix, resulting in a stronger material. At high levels of crystallinity, microcracking appears to be an effective toughening mechanism.

3. For the lithium disilicate glass/glass-ceramic system, the stress corrosion susceptibility (N) is more dependent on aging in water than on the percent of crystallinity. The kinetics of aqueous corrosion of the residual glassy phase and any residual stresses present play important roles in determining the fatigue behavior of all materials tested.

4. For slow crack growth experiments, the interactions of the crack front with the microstructure greatly influence the determination of the V-K relationship. A method such as the constant moment double cantilever beam technique wherein the crack growth is visually monitored is ideal for

monitoring this interaction. Marking the locations of features of interest combined with subsequent fractographic analysis should provide a powerful tool in studying crack/microstructure interactions.

5. The N values obtained in crack growth studies more closely correspond to the lowest values obtained in dynamic fatigue. This is likely due to the kinetics of aqueous corrosion governing the growth of the macroscopic stress state controlling the fracture.

6. The empirical relationships used in these analyses for lifetime prediction provide inconsistent results for materials tested under variable conditions. The theory needs to be extended to account for a change in the flaw configuration and surface composition if reliable predictive capabilities are to be achieved.

## ACKNOWLEDGEMENTS:

The authors gratefully acknowledge partial financial support of AFOSR Contract # F49620-80-C-0047.

The authors also acknowledge helpful discussions with D. Greenspan and S. Freiman during this work.

REFERENCES:

1. S.M. Wiederhorn, "Subcritical Crack Growth in Ceramics," pp. 613-646 in Fracture Mechanics of Ceramics, Vol. 2, R.C. Bradt, D.P.H. Hasselman, and F.F. Lange, editors, Plenum Press, New York, 1974.

2. J.E. Ritter, Jr., D.C. Greenspan, R.A. Palmer, and L.L. Hench, "Use of Fracture Mechanics Theory in Lifetime Predictions for Alumina and Bioglass-Coated Alumina," J. Biomed. Mater. Res., 13, pp. 251-263, 1979.

3. J.E. Ritter, Jr., "Engineering Design and Fatigue Failure of Brittle Materials," pp. 667-686 in Fracture Mechanics of Ceramics, Vol. 4, R.C. Bradt, D.P.H. Hasselman, and F.F. Lange, editors, Plenum Press, New York, 1978.

4. D.C. Greenspan, "The Chemical, Mechanical, and Implant Properties of Glass-Coated Alumina," Ph.D. Dissertation, University of Florida, Gainesville, Florida, 1977.

5. R.A. Palmer, Fracture Mechanics and Failure Predictions for Glass, Glass-Ceramic, and Ceramic Systems," Ph.D. Dissertation, University of Florida, Gainesville, Florida, 1981.

6. S.W. Freiman, "The Relation of the Kinetics of Crystallization to the Mechanical Properties of $LiO_2$-$SiO_2$ Glass-Ceramics," Ph.D. Dissertation, University of Florida, Gainesville, Florida, 1968.

7. R.T. DeHoff and F.N. Rhines, Quantitative Microscopy, McGraw-Hill, New York, 1968.

8. J.B. Wachtman, W. Capps, and J. Mandel, "Biaxial Flexure Testing of Ceramic Substrates," J. Mater. 7(2), pp. 188-194, 1972.

9. S.W. Freiman, D.R. Mulville, and P.W. Mast, "Crack Propagation Studies in Brittle Materials," J. Mater. Sci. 8(11) pp. 1527-1533, 1973.

10. W. McCracken, D.E. Clark, and L.L. Hench, "Aqueous Durability of Lithium Disilicate Glass-Ceramics, Am. Ceram. Soc. Bull., 61(11) 1218-1223, 1982.

11. S.W. Freiman and L.L. Hench, "Effect of Crystallization on the Mechanical Properties of $Li_2O$-$SiO_2$ Glass-Ceramics," J. Am. Ceram. Soc., 55(2) 86-90, 1972.

12. D.P.H. Hasselman and R.M. Fulrath, "Proposed Fracture Theory of a Dispersion-Strengthened Glass Matrix," J. Am. Ceram. Soc. 49(2), pp. 68-72, 1966.

13. N. Miyata and H. Jinno, "Theoretical Approach to the Fracture of Two-Phase Glass-Crystal Composites," J. Mater. Sci. 7(9), pp. 973-982, 1972.

14. A.S. Rao, Ph.D. Dissertation, University of British Colombia, Vancouver, 1977.

15. K.T. Faber and A.G. Evans, "Toughening in Lithium-Aluminosilicate Glass-Ceramics," Paper #71-B-80P at the 1980 Pacific Coast Section Meeting of the American Ceramic Society, Abstract in Cer. Bull. 59(8), p. 829, 1980.

16. L.L. Hench and D.E. Clark, "Physical Chemistry of Glass Surfaces," J. Non-Cryst. Solids 25, pp. 343-369, 1977.

17. D.M. Sanders and L.L. Hench, "Mechanisms of Glass Corrosion," J. Am. Ceram. Soc. 56(7), pp. 373-377, 1973.

18. B,J. Pletka and S.M. Wiederhorn, "Subcritical Crack Growth in Glass-Ceramics," pp. 745-759 in Fracture Mechanics of Ceramics, Vol. 4, R.C. Bradt, D.P.H. Hasselman, and F.F. Lange, editors, Plenum Press, New York, 1978.

19. P.S. Theocaris and J. Milios, "Crack Propagation Velocities in Bi-phase Plates Under Static and Dynamic Loading," Eng. Fract. Mech. 13(3), pp. 599-610, 1980.

20. M.K. Ferber and S.D. Brown, "Subcritical Crack Growth in Dense Alumina Exposed to Physiological Media," J. Am. Ceram. Soc. 63(7), pp. 424-429, 1980.

21. E.R. Fuller, Jr., "An Evaluation of Double-Torsion Testing - Analysis," pp. 3-18 in Fracture Mechanics Applied to Brittle Materials, ASTM STP 678, S.W. Freiman, editor, American Society for Testing and Materials, Philadelphia, PA, 1979.

22. B.J. Pletka, E.R. Fuller, Jr., and B.G Koepke, "An Evaluation of Double-Torsion Testing - Experimental," pp. 19-37 in Fracture Mechanics Applied to Brittle Materials, S.W. Freiman, editor, American Society for Testing and Materials, Philadelphia, PA, 1979.

Table I. Fracture Parameters from Dynamic Fatigue Experiments

| | N | lnB |
|---|---|---|
| 33L-Glass | | |
| Air | 30.5 | -2.07 |
| Water | 11.0 | 4.23 |
| Aged 1 Day | 37.4 | -30.73 |
| 33L-10% Crystalline | | |
| Air | 116.1 | -34.43 |
| Water | 136.0 | -53.93 |
| Aged 1 Day | 46.3 | -14.88 |
| Aged 1 Week | 27.4 | -1.13 |
| 33L-60% Crystalline | | |
| Aged 1 Day | 59.2 | -10.16 |
| Aged 1 Week | 23.6 | 0.06 |
| 33L-92% Crystalline | | |
| Air | 70.6 | -5.66 |
| Water | 25.7 | 2.10 |
| Aged 1 Day | 27.6 | 0.73 |
| Aged 1 Week | 22.3 | 2.37 |

Table II. Strength of 33L-Glass and Glass-Ceramics

| | Inert Strength* | Ambient Strength** |
|---|---|---|
| Glass | | |
| As prepared (Air) | 179 | 125 |
| Water | --- | 112 |
| Aged 1 Day | 211 | 129 |
| 33L-92% Crystalline | | |
| As prepared (Air) | 274 | 226 |
| Water | --- | 199 |
| Aged 1 Day | 280 | 203 |
| Aged 1 Week | 221 | 160 |
| 33L-10% Crystalline | | |
| As prepared (Air) | 233 | 155 |
| Water | --- | 141 |
| Aged 1 Day | --- | 136 |
| Aged 1 Week | --- | 153 |
| 33L-60% Crystalline | | |
| Aged 1 Day | 210 | 148 |
| Aged 1 Week | 216 | 131 |

*Average strength (MPa), tested in liquid nitrogen at 16 $MPa\text{-}s^{-1}$

**Average strength (MPa), tested at room temperature at 16 $MPa\text{-}s^{-1}$

Table III. Fracture Parameters from Slow Crack Growth Experiments

| | N | ln A |
|---|---|---|
| 33L Glass | 11.0 | -157.35 |
| 33L-10% Crystalline | 12.1 | -183.4 |
| 33L-60% Crystalline | 15.3 | -232.6 |
| 33L-92% Crystalline | 14.4 | -218.7 |

Table IV. Proof Test Stress Ratios Derived from Dynamic Fatigue and Slow Crack Growth Experiments

| | Dynamic Fatigue $(\sigma_p/\sigma_a)$* | Slow Crack Growth $(\sigma_p/\sigma_a)$* |
|---|---|---|
| 33L-Glass | | |
| Air | 2.8 | 0.63 |
| Water | 14.1 | ---- |
| Aged 1 Day | 5.5 | ---- |
| 33L-10% Crystalline | | |
| Air | 1.8 | 0.37 |
| Water | 1.9 | ---- |
| Aged 1 Day | 2.6 | ---- |
| Aged 1 Week | 3.0 | ---- |
| 33L-60% Crystalline | | |
| Aged 1 Day | 1.9 | 0.84 (Air) |
| Aged 1 Week | 3.4 | ---- |
| 33L-92% Crystalline | | |
| Air | 1.7 | 1.05 |
| Water | 3.0 | ---- |
| Aged 1 Day | 2.9 | ---- |
| Aged 1 Week | 3.4 | ---- |

*For 10 years at 50 MPa

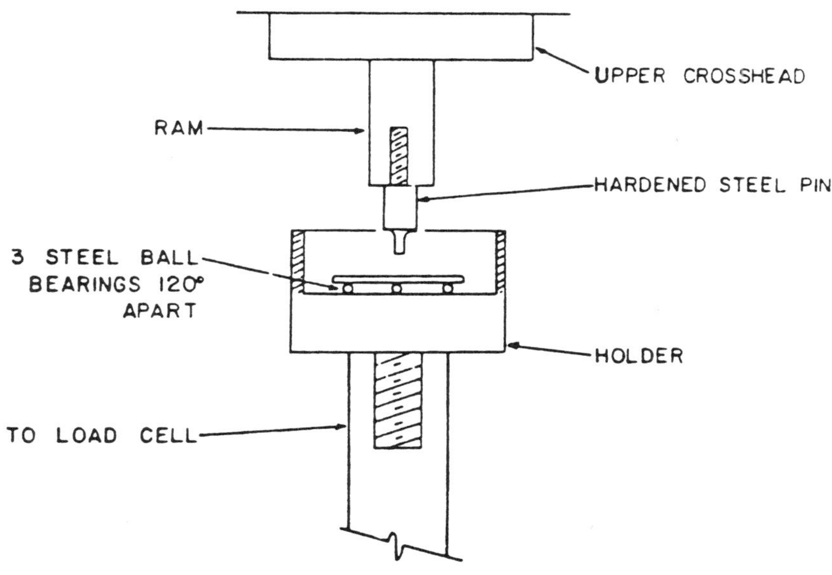


Figure 1. Biaxial flexure test jig

Figure 2. Constant moment double cantilever beam test apparatus

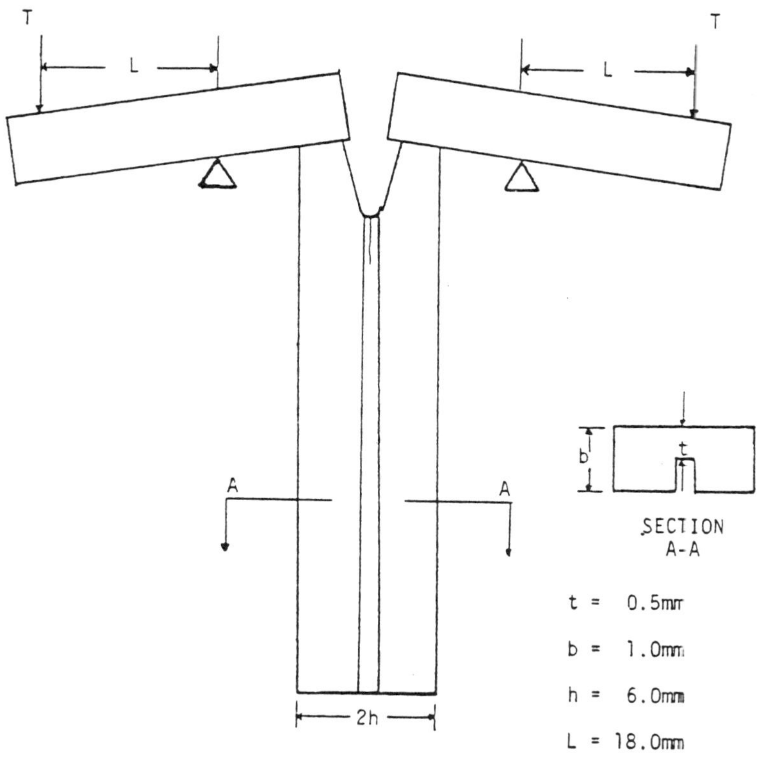


Figure 3. Specimen configuration for double cantilever beam constant moment crack growth experiment

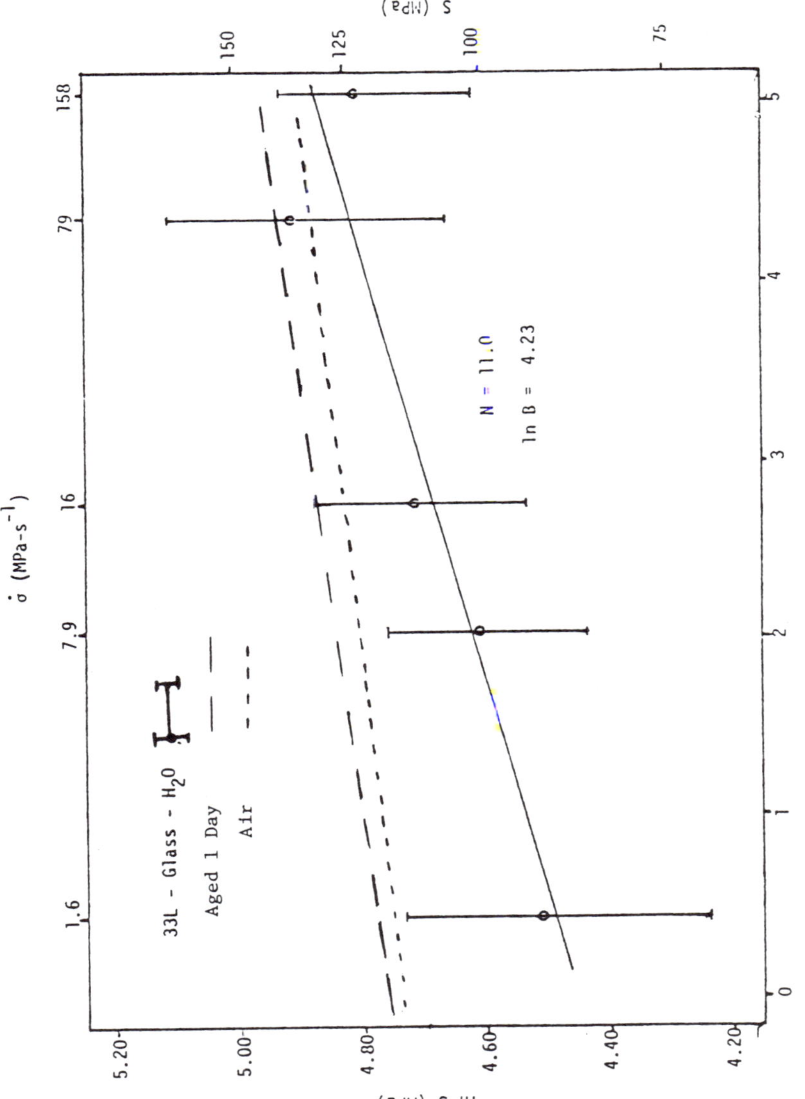


Fig. 4. Dynamic fatigue results fro 33L glass in different environmental test conditions.

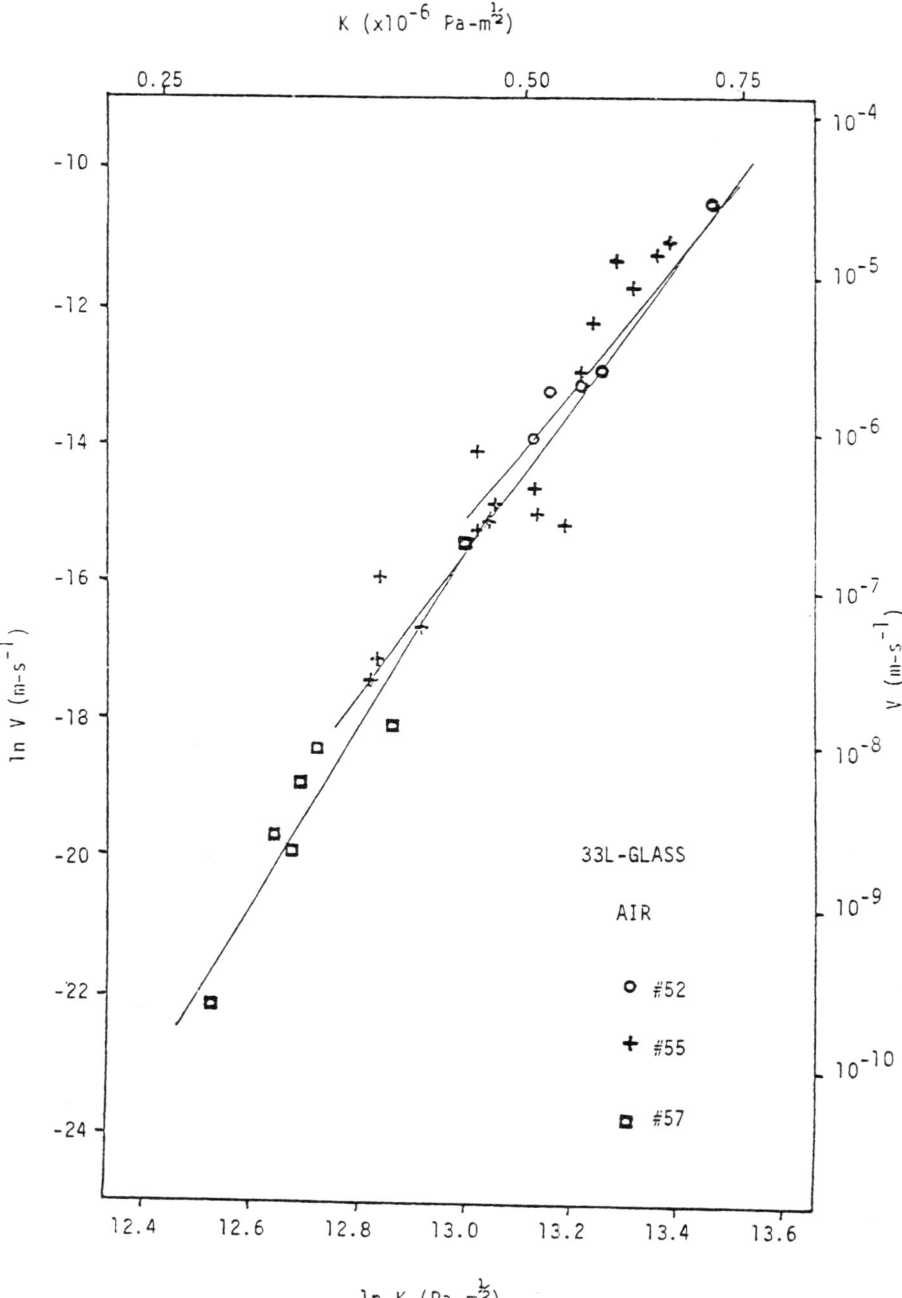


Figure 5A. V-K relationship for 33L-Glass tested in air

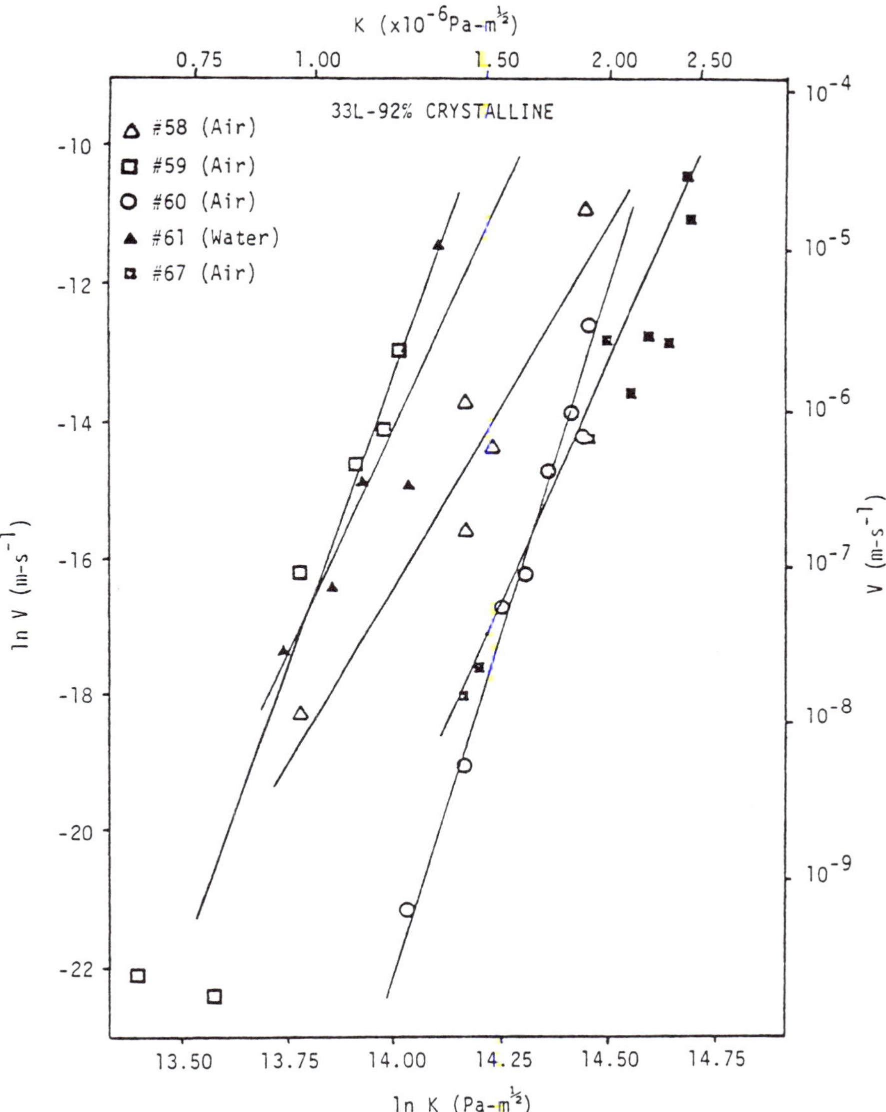


Figure 5B. V-K relationship for 33L-92% crystalline

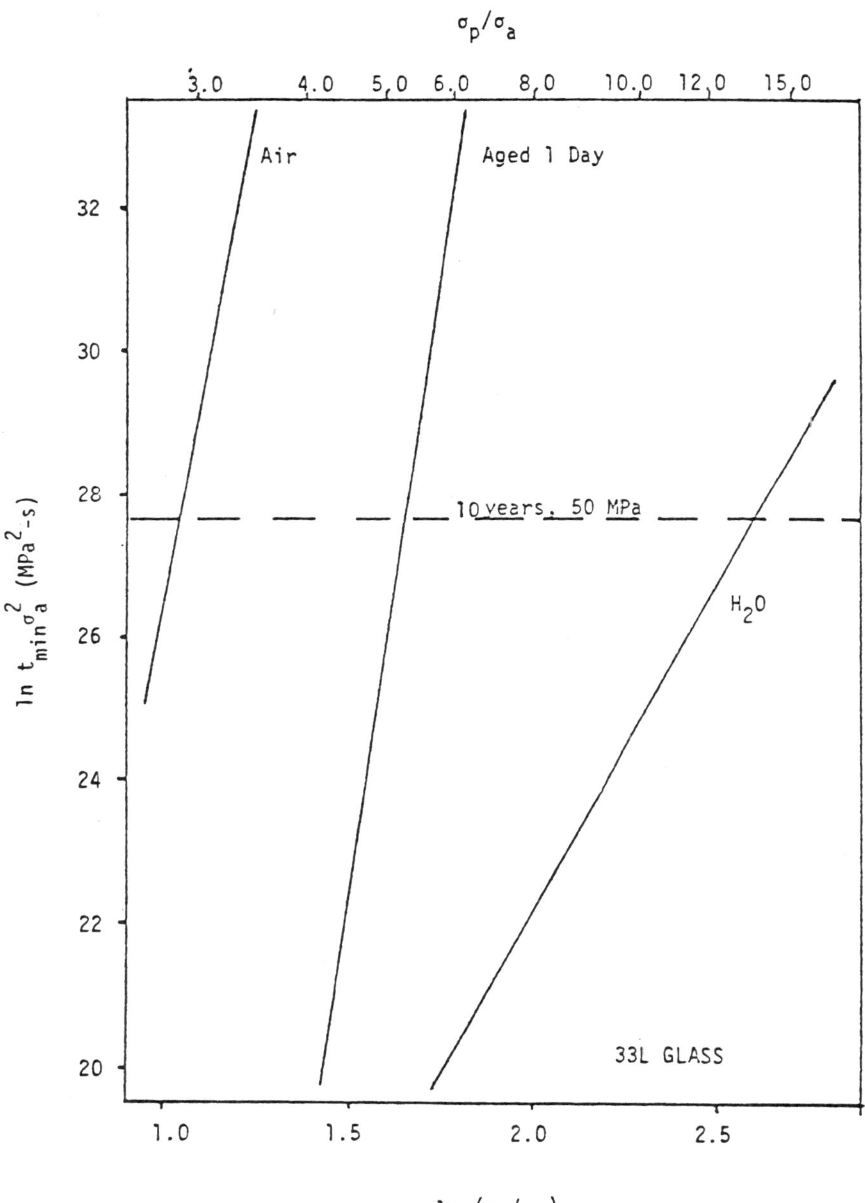


Figure 6A. LPD for 33L-Glass

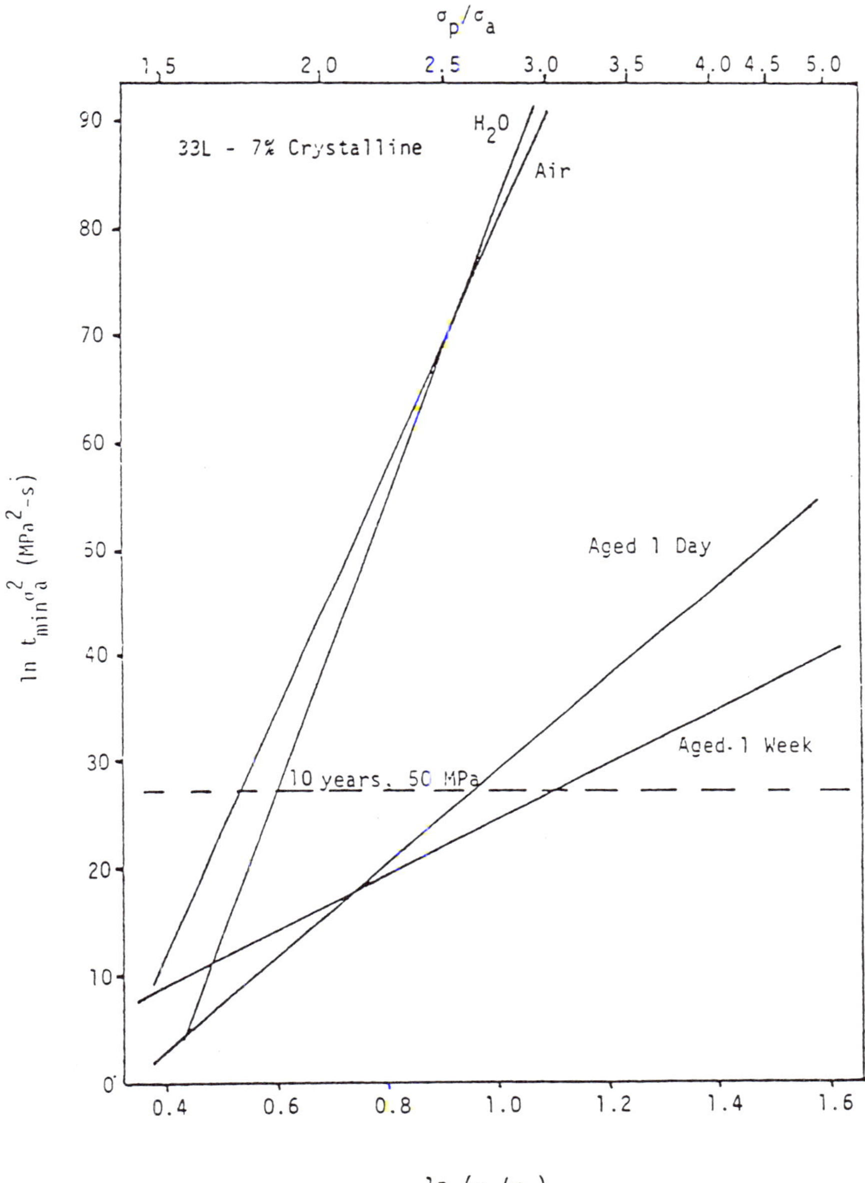


Figure 6B. LPD for 33L-7% crystalline

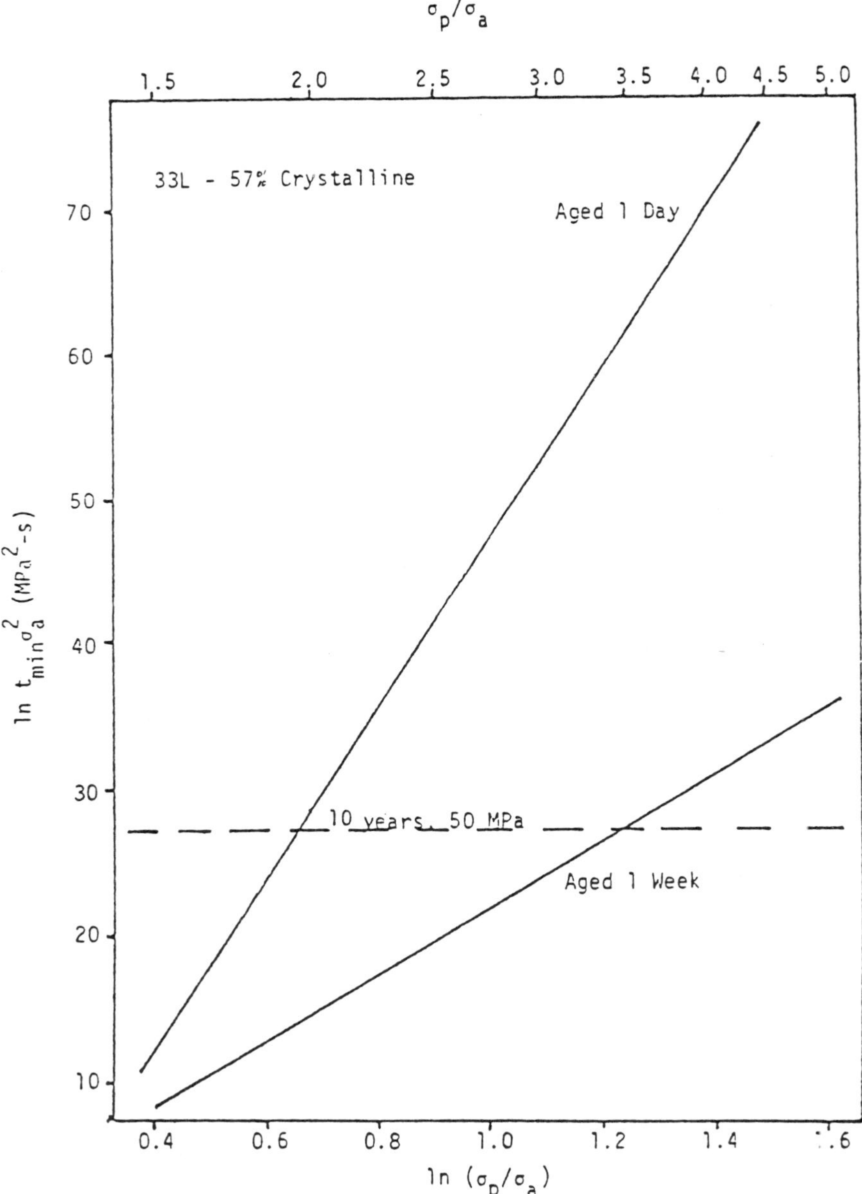


Figure 6C. LPD for 33L-57% crystalline

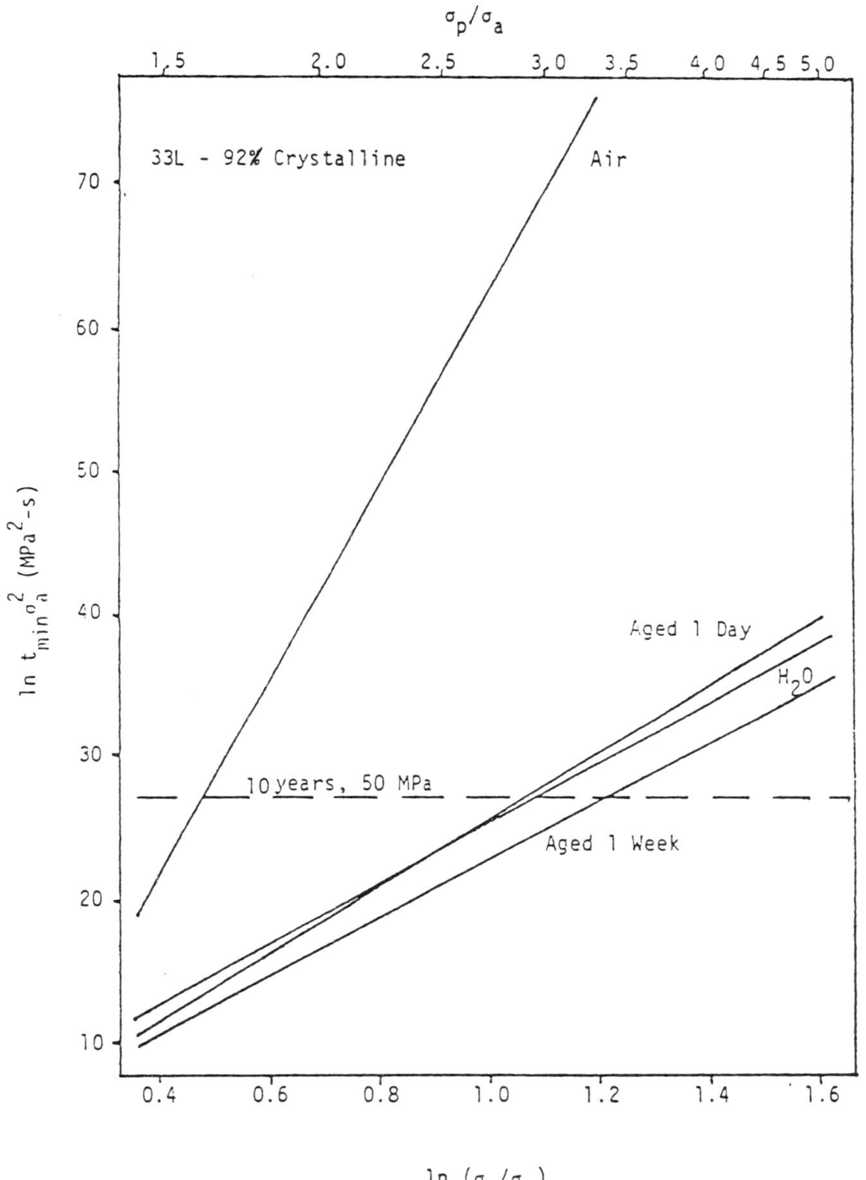


Figure 6D. LPD for 33L-92% crystalline

# PREPARATION OF $xNa_2O$-(1 - x)$SiO_2$ GELS FOR THE GEL-GLASS PROCESS: I. ATMOSPHERIC EFFECT ON THE STRUCTURAL EVOLUTION OF THE GELS

M. Prassas, J. Phalippou, L.L. Hench and J. Zarzycki

*Laboratory of Materials Science and CNRS Glass Laboratory*
*University of Montpellier II, France*

The gels of the binary system $Na_2O-SiO_2$ are difficult to prepare under reproducible conditions because of the catalytic effect of the $OH^-$ ions. The gels were prepared in the whole range of composition which corresponds to the vitrification range. The structural analogy between the gels and the corresponding glasses was shown by X-ray diffraction and IR spectroscopy. The problems related to the preparation of these gels were studied, following their structural evolution during ageing. The action of atmospheric agents ($CO_2$ and $H_2O$) induces a crystallization of carbonate compounds. The crystalline phase which appears is $Na_3H(CO_3)_2 2 H_2O$ in the case of gels containing less than 25 mol% $Na_2O$. For a higher sodium content the compound which crystallizes is $Na_2CO_3-H_2O$. The interaction between the atmosphere and the gel starts at the "surface" of the samples as observed by the IR reflection spectroscopy.

## 1. Introduction

Different unconventional methods can be used to obtain vitreous materials [1]. Among these the gel route is of great interest as has already been shown [2–7].

Oxide glasses which are difficult to prepare by the classical method of fusion, e.g. $SiO_2$, $SiO_2-TiO_2$, $SiO_2-B_2O_3$, $SiO_2-P_2O_5$, can be obtained with high silica content by hot-pressing the corresponding gels [8–10].

One of the main advantages of the gel route is the synthesis of very homogeneous materials.

Not much interest has been shown until now in the reactions at low temperatures, e.g. during ageing which may influence the ultimate homogeneity of the material.

The present work was undertaken to clarify the role of the atmosphere in the structural evolution of the gels in the binary system $xNa_2O-(1-x)SiO_2$.

The structural evolution as a function of heat treatment and the gel–glass transformation of this system will be presented in a separate paper.

## 2. Preparation of gels

### *2.1. Starting materials*

The gels were synthesized by hydrolysis and polycondensation techniques of organometallic compounds (MOD * method [11]).

Tetramethoxysilane (TMS **) was used as a source of $SiO_2$ and sodium methylate † diluted in methanol (30% in volume) for $Na_2O$. The gelification was obtained in the presence of methanol †† which is soluble with two organometallics and the distilled water necessary for hydrolysis.

### *2.2. Gelation process*

In the basic medium the polycondensation reaction of the polysilicic acids is very rapid [12]. This is true for the case of the mixture $CH_3OH–Si(OCH_3)_4–NaOCH_3–H_2O$ according to the classical hydrolysis (1) and polycondensation (2) reactions:

$$Si(OCH_3)_4 + 4\,H_2O \rightarrow Si(OH)_4 + 4\,CH_3OH \tag{1}$$

$$\equiv Si–OH + OH–Si\equiv \rightarrow \equiv Si–O–Si\equiv + H_2O \tag{2}$$

together with the hydrolysis reaction of $NaOCH_3$:

$$NaOCH_3 + H_2O \rightarrow Na^+ + OH^- + CH_3OH. \tag{3}$$

It is sometimes impossible to add the stoichiometric quantity of water for the hydrolysis of TMS; the presence of $OH^-$ ions according to eq. (3) provokes a rapid gelation, which cannot be controlled.

For these reasons, all the preparations were made at 0°C. The mixture were prepared with 50% of organometallic derivatives in methanol. The quantity of water used for hydrolysis was chosen so that the gelation of the mixture was completed in about 5 to 10 min.

For these experimental conditions, the quantity of water necessary for hydrolysis depends on the $NaOCH_3$ content (fig. 1a). However, it does never exceed 4 mol $H_2O$/mol TMS. It appears that the $\equiv Si–O–Si\equiv$ bonds formed during reaction (2) can break because of the depolymerization role of the $Na^+$ ions according to reaction (4).

$$\equiv Si–O–Si\equiv + Na^+ + OH^- \rightarrow \equiv Si–O^-Na^+ + \equiv Si–OH. \tag{4}$$

The role played by the $Na^+$ ions, which is well known in the binary system $SiO_2–Na_2O$ is to decrease the number of bridging oxygens when the $Na_2O$ content increases (fig. 1b). The process of gelation would finally depend on the reaction (1) and (4).

* Metallic-organic derived (MOD).

** $Si(OCH_3)_4$: Fluka product, purum quality.

† $NaOCH_3$: Fluka product, parct. quality.

†† $CH_3OH$: Prolabo product, Normapur quality.

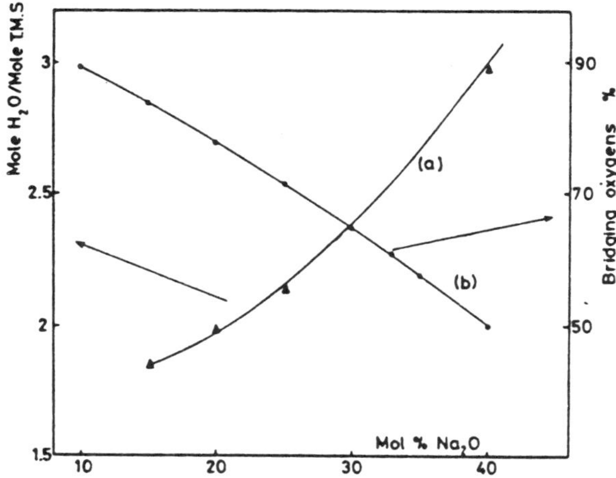


Fig. 1. The amount of water for hydrolysis (a) and percentage of bridging oxygens (b) versus the $Na_2O$ content.

To obtain samples of suitable dimensions the drying was preceded by an ageing of the gel in the "parent solution" (solvent obtained from syneresis) at 55°C for 5 h.

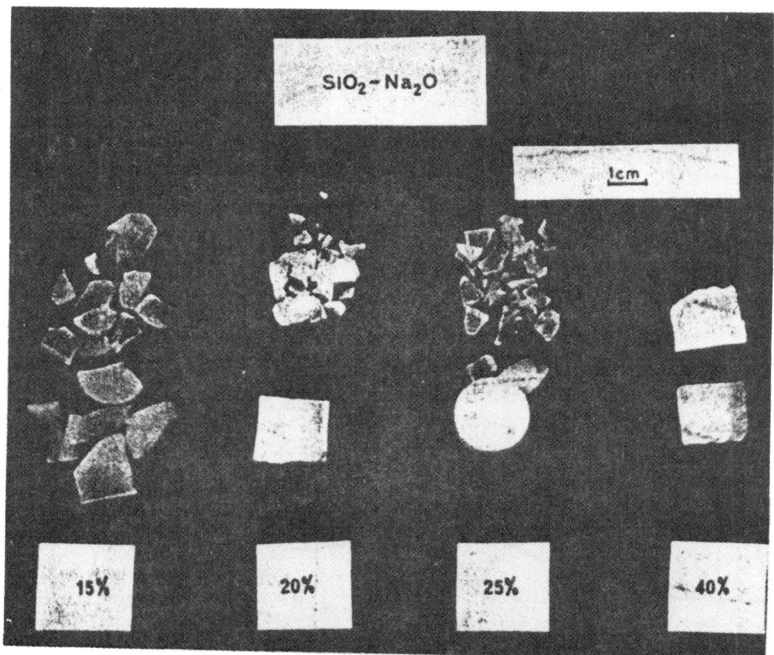


Fig. 2. Dried gels.

The gels were then dried for 10 h at 130°C under methanolic atmosphere. After drying, the gels were immediately placed in a dessicator and conserved under a primary vacuum.

The gels thus obtained (fig. 2) are represented by "aN", where "a" denotes the molar percentage of $Na_2O$.

## 3. Results and discussion

### *3.1. Structural characterization of the gels*

This was essentially made by three different techniques: X-ray diffraction, IR spectroscopy in transmission and reflection.

#### *3.1.1. X-ray diffraction*

The structural evolution of the gels was followed by X-ray diffraction using a diffractometer (Philips 1140) with a proportional counter. $Cuk_\alpha$ radiation was used with a Ni filter.

The samples were prepared in the form of a powder having the same grain size ($< 60\ \mu$) and placed in a flat identical support.

Fig. 3 shows the X-ray diffraction spectra obtained for the freshly prepared samples (immediately after the drying).

Their common characteristic is the presence of a "diffuse pattern" indicating the amorphous structure of the gels.

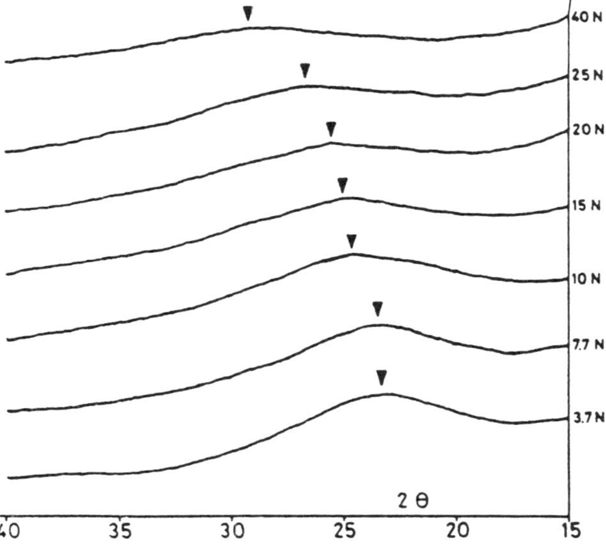


Fig. 3. X-ray diffraction spectra of the gels $x\,Na_2O-(1-x)SiO_2$.

The analysis of these spectra shows that the maximum of the diffuse pattern is displaced towards a higher angle of diffraction with the increase of the $Na_2O$ content. This shift is of the order of $2\theta = 6°$ between the gel containing the least amount of $Na_2O$ (3.7 N) and the gel 40 N. This change is considered significant and it can probably be related to the structural change occurring in these gels due to the addition of $Na^+$ ions.

*3.1.2. IR transmission spectroscopy*

For the IR absorption spectra, the samples were prepared using the KBr pellets technique. The spectra were recorded in a double beam spectrometer (Perkin-Elmer 577). A pure KBr pellet was placed on the reference beam.

The IR transmission spectra of the gels $SiO_2$, 20 N and 40 N are shown in fig. 4. The silica gel, included in the present study for comparison, was prepared by the MOD method. In the spectrum of the silica gel, three characteristic absorption bands of amorphous silica are observed between 400 and 1300 $cm^{-1}$.

The (S) band at 1080 $cm^{-1}$ is attributed to the stretching vibration of the Si–O bands [13]. The (D) band at 455 $cm^{-1}$ is attributed to the vibrational modes of deformation of the O–Si–O and Si–O–Si bonds [13,14] and the (R)

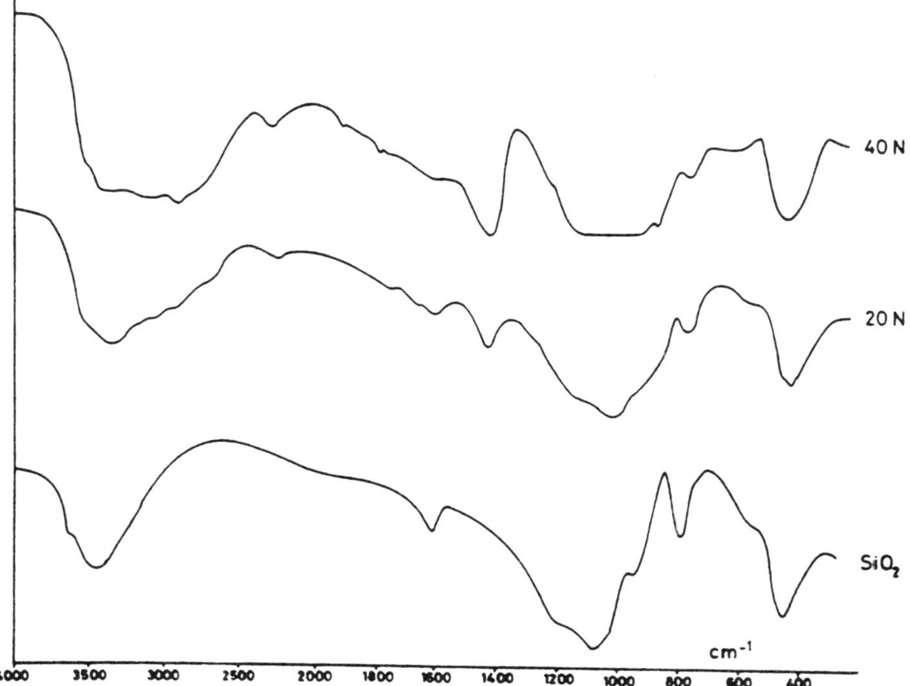


Fig. 4. IR transmission spectra (KBr pellet technique) of the gels $SiO_2$, 20N and 40N.

band located at 800 $cm^{-1}$ is attributed to the ring structure of the tetrahedra ($SiO_4$) [5]. Lippencott et al. [13] attributed the latter to the stretching mode of the ← Si–O–Si → .

Three other bands appear on the IR transmission spectra of the silica gel at 3500, 1620 and 950 $cm^{-1}$. The bands at 3500 and 1620 $cm^{-1}$ are attributed respectively to the stretching and deformation modes for hydroxyl groups and molecular water [16].

The band at 950 $cm^{-1}$ is usually attributed to the vibration of the $Si–O^-$ bonds for the glasses containing non-bridging oxygens [7]. In the silica gel it is preferable to attribute this band to the stretching mode of the Si–O (OH) bonds [18]. This is in good agreement with the evolution of this band during the densification treatment of the gel [8]; it disappears at higher temperature.

In the case of the gel 20N, the (S) and (R) bands are shifted by 50 and 20 $cm^{-1}$ respectively towards lower wavenumbers with respect to those of pure silica gel (fig. 4).

A similar shift observed in the binary glasses $SiO_2–Na_2O$ [17,19] is related to a decrease of the force constant of the bonds between the Si atom and non-bridging oxygen [17].

The intensity of the (R) band decreases when the $Na_2O$ content increases up to 40 mol%. This effect indicates the depolymerizing role of the $Na^+$ ions in the gels. The probability of a ring arrangement of tetrahedra becomes progressively lower for the gels containing a higher amount of $Na_2O$.

A considerable modification of the band at 3500 $cm^{-1}$ is observed with an increase of $Na_2O$ content. In fact, the intensity of this band increases and becomes wider in the region of lower wavenumbers. The interaction between silanol group, $Si–O^-$ and adsorbed water by hydrogen bonds are responsible for the observed widening.

The IR band at 2800 $cm^{-1}$, the intensity of which increases as a function of $Na_2O$ content, is attributed to the vibrations of the OH groupings strongly linked by the hydrogen bonds.

Finally, it should be noted that there is an absorption band located at 1450 $cm^{-1}$ which is not observed in the case of silica gel. The intensity of this band depends on the concentration of $Na_2O$ (fig. 5).

The IR spectra of the binary glasses $SiO_2–Na_2O$ also show a band at 1430 $cm^{-1}$, but this was not attributed to any particular vibration [17,20]. This is certainly the consequence of the attack on the material by atmospheric agents. Indeed, for the sample 20N, the increase of the band at 1450 $cm^{-1}$ as a function of exposure time in air shows that its presence depends on the "history" of the gel (fig. 6). It is a band characteristic of the sodium carbonate as shall be shown later.

*3.1.3. IR reflection spectroscopy*

In materials containing alkali ions, the study of IR transmission spectra by the KBr technique can give rise to erroneous results. Indeed, it was shown that there is a possibility of a reaction between the KBr and the gel [21]. For this

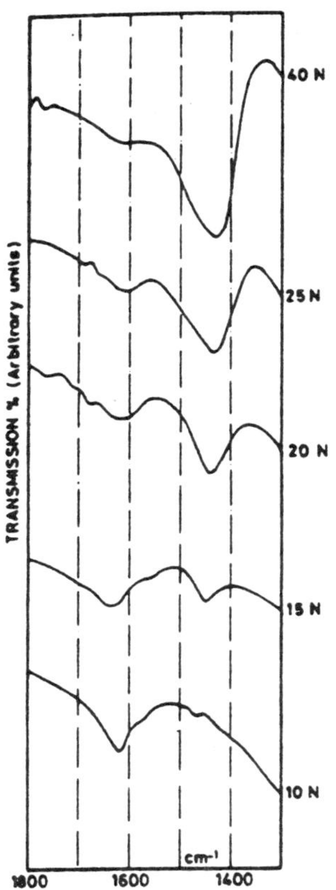


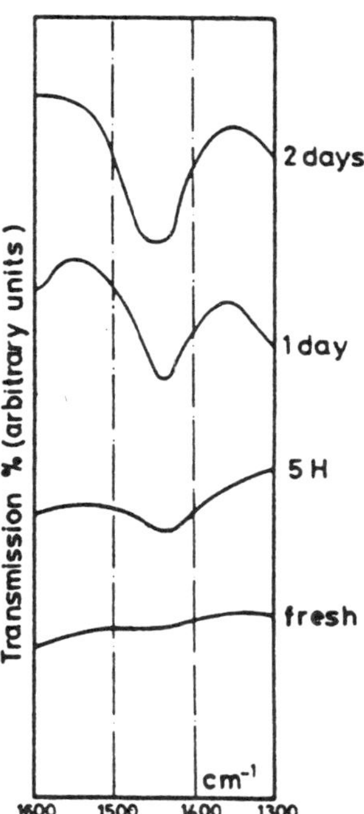


Fig. 5. Evolution of the IR band located at 1450 $cm^{-1}$ as a function of the $Na_2O$ content.

Fig. 6. Evolution of the IR band located at 1450 $cm^{-1}$ in relation to exposure time in air for the gel 20N.

reason, it seemed important to use the IR reflection technique, which has recently been used for the study of the gels to follow the evolution of the gel–glass transformation [22].

The IR reflection spectra were recorded on the same spectrometer. All the gels were polished using emery paper (240–400–600–1200). The intensity of the reference beam was chosen at 1150 $cm^{-1}$ and corresponds to 80% of the reflection from a silica glass polished using emery paper 600.

The reflection spectra of the gels are shown in fig. 7. The positions of the principal bands located between 1300 and 200 $cm^{-1}$ are in good agreement with those obtained in transmission spectra. In a silica gel the (S), (R) and (D) bands are located at 1070, 775 and 440 $cm^{-1}$, respectively. With respect to the same bands in transmission, they are all shifted from 10 to 25 $cm^{-1}$ towards the lower wavenumber. The increase of $Na_2O$ content simultaneously induces

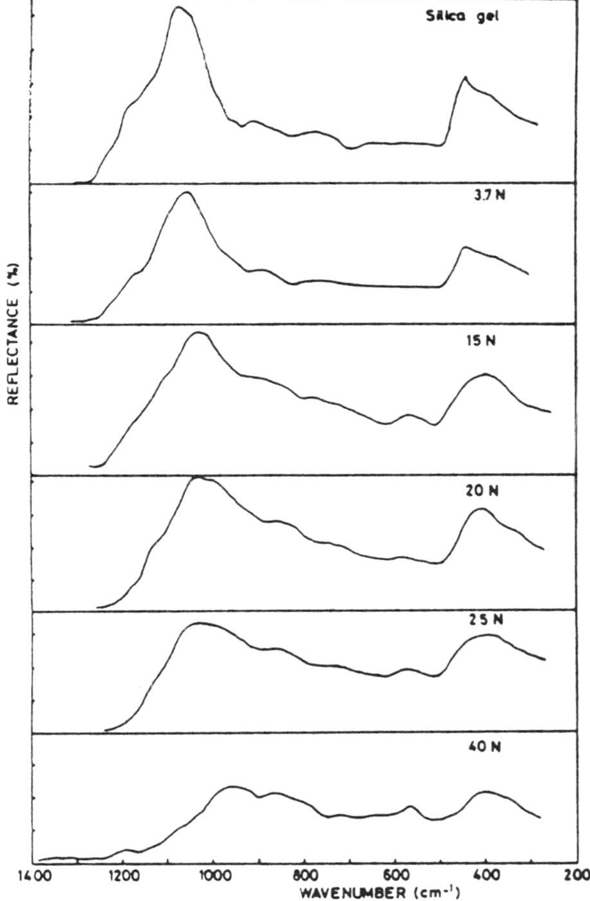


Fig. 7. IR reflection spectra of the gels $x Na_2O-(1-x)SiO_2$.

a widening of the "band" centered around 1070 $cm^{-1}$ and a shift towards the lower wavenumber.

Fig. 8 shows the evolution of the (S), (D) and (R) bands and of the band due to the non-bridging oxygens as a function of $Na_2O$ content. The same evolution has already been observed and discussed for glasses of similar composition [19]. A structural similarity between the gels and glasses follows from this observation.

### *3.2. Behaviour of the gels during the ageing in air*

In the study of the IR spectra of these gels, it was observed that spectra were altered by the exposure of the samples to air. To understand this

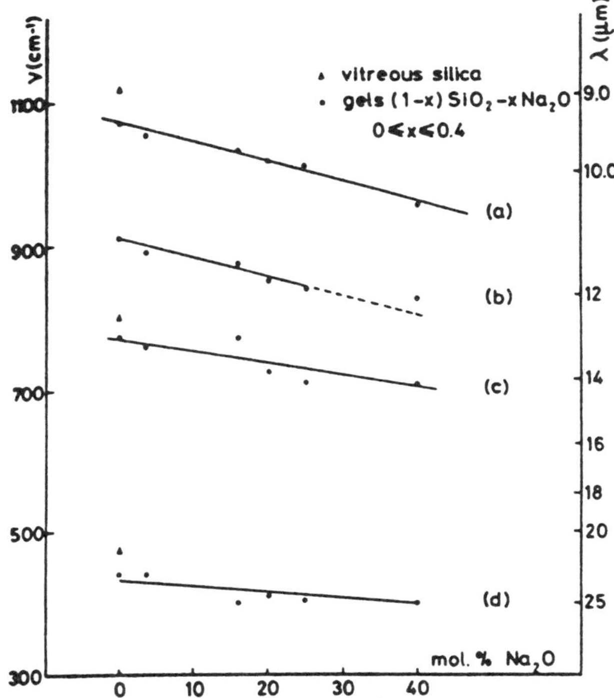


Fig. 8. Frequency of the IRRS band in relation to the molar percentage of $Na_2O$: (a) stretching vibration Si–O; (b) stretching vibration $Si-O^-$; (c) vibration due to the ring structure ($SiO_4$); (d) deformation vibration O–Si–O and Si–O–Si.

phenomenon the samples were studied by X-ray diffraction, as a function of their ageing in air. The humidity and the average temperature were 50% RH and 25°C respectively.

In the gel 15 N (fig. 9), a crystalline phase appears, the most intense peak of which corresponds to a *d*-spacing of 2.63 Å. This is observed after an exposure time of 20 h. The intensity of this peak increases progressively as a function of time. Other peaks appear simultaneously, which indicates an increase in the crystalline phase.

The spectra do not reveal any significant modification after 18 days exposure in air. The nature of the crystalline phase was identified (ASTM) as a mixed carbonate $Na_3H(CO_3)_2 2H_2O$, commonly called trona.

The appearance of crystals in the gel 20N is more rapid than in the gel 15 N (fig. 10). Three peaks corresponding to *d*-spacing 2.77–2.66 and 2.36 Å emerge from the diffuse pattern after 20 h exposure in air. Their evolution during the ageing of the gels in air shows that probably two kinds of crystals are present initially: the thermonatrite $Na_2CO_3H_2O$ and trona. The latter increases at the expense of the $Na_2CO_3H_2O$. At the end of the treatments only the compound trona is observed on the spectra.

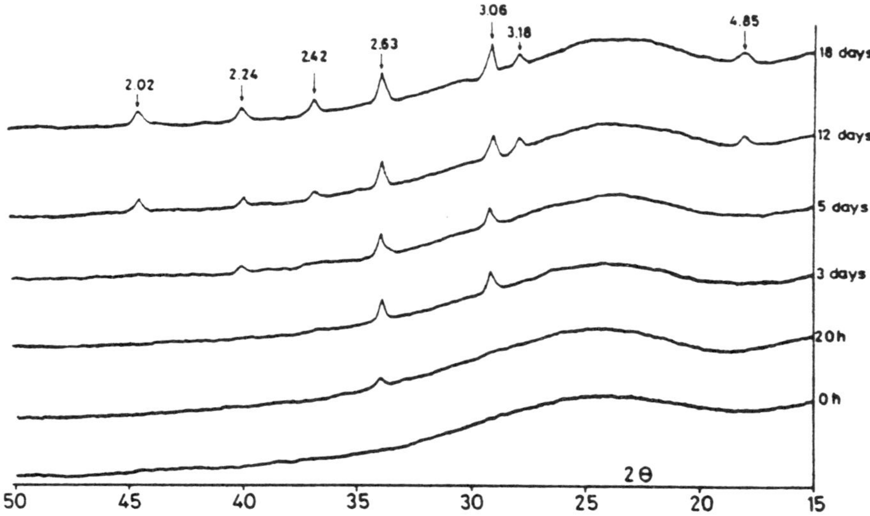


Fig. 9. X-ray diffraction spectra as a function of the exposure time in air for the gel 15N.

The crystalline phase initially present in the gel 25N (fig. 11) is thermonatrite. After 3 days exposure in air, the spectra of this gel revealed the most intense peaks of the compound trona. This carbonate increases during the

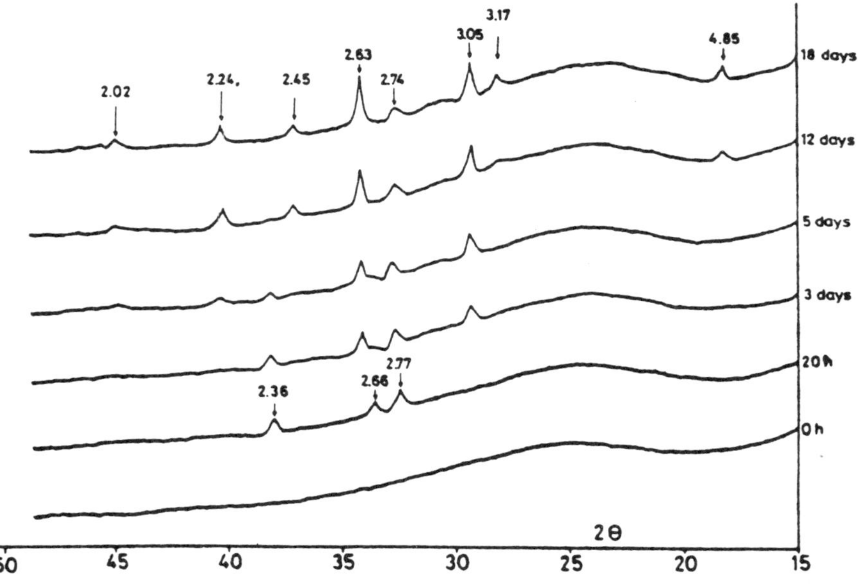


Fig. 10. X-ray diffraction spectra as a function of the exposure time in air for the gel 20N.

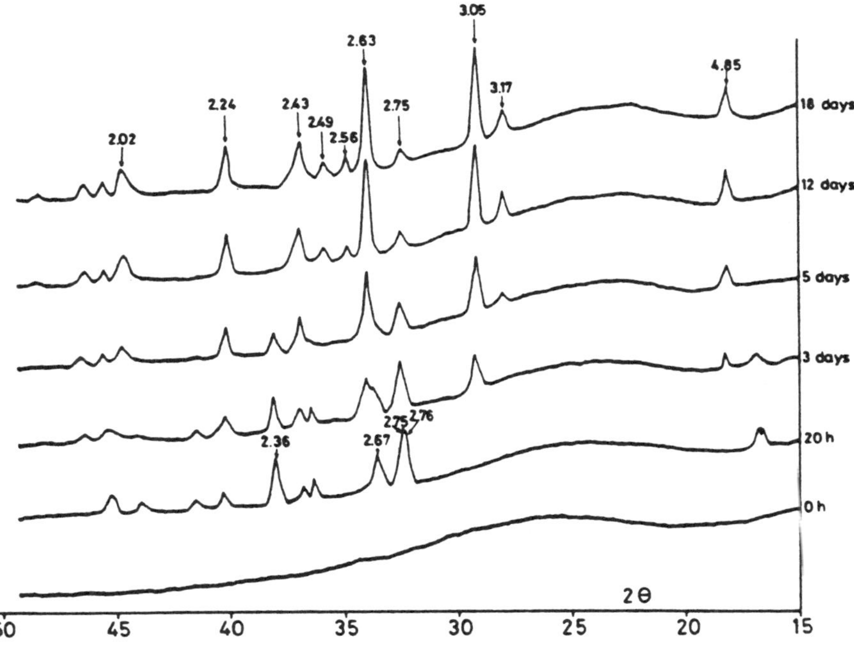


Fig. 11. X-ray diffraction spectra as a function of the exposure time in air for the gel 25N.

ageing process at the expense of thermonatrite, which progressively disappears.

Finally, in the gel 40 N the carbonate which crystallizes is $Na_2CO_3$ (fig. 12). The amount of Na-carbonate increases regularly as a function of ageing treatment. This compound becomes hydrated (fig. 12).

In contrast to the other gels, the compound trona does not appear even after an exposure time of one month. It is noted that the behaviour of the gel 40N is altogether different from that of the other gels. Indeed the probable attack by $CO_2$ and the water vapour present in air induces the appearance of a rigid surface layer which is white in colour. The inside of the sample becomes rapidly paste-like as a result of this attack. It is known that glasses of the same composition are extremely soluble in water [23]. Thus it seems that the attack by air, which creates a layer of carbonate on the surface, is followed by the dissolution of the gel by the water present in atmosphere.

Table 1 shows all the crystalline phases formed in these gels studied at the beginning and end of the atmospheric attack (one month). The results concerning the gels 10N and 33N are also included.

### 3.3. *Localization of the crystals of carbonates on the surface of the gels*

The morphology of the crystals of sodium carbonates can be studied by scanning electron microscopy. The micrographs shown in figs. 13(a) and 13(c)

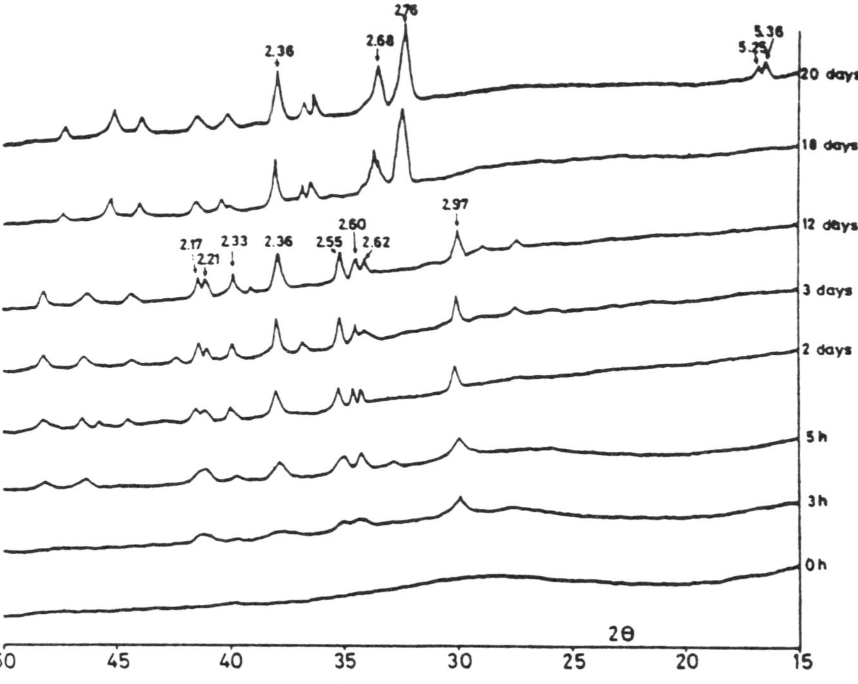


Fig. 12. X-ray diffraction spectra as a function of the exposure time in air for the gel 40N.

indicate the different surface morphologies of the samples 15N and 40N respectively, having been aged in air for 2 days.

The surface of the gel 15N, which was kept in the dessicator for one month, does not display any crystalline structure except in some parts where the crystallization has just started [white colored grains can be seen in the fig. 13(b)].

Observation of a freshly fractured sample (40N), aged for 2 days, indicates

Table 1
Various Na-carbonates formed in gels having different sodium contents after ageing in air

| $x$ | Crystalline phase after 20 h | Crystalline phase after one month |
|---|---|---|
| 0.10–0.15 | trona | trona |
| 0.20 | thermonatrite + trona | trona |
| 0.25 | thermonatrite | trona |
| 0.33 | thermonatrite | thermonatrite |
| 0.40 | Na-carbonate | thermonatrite |

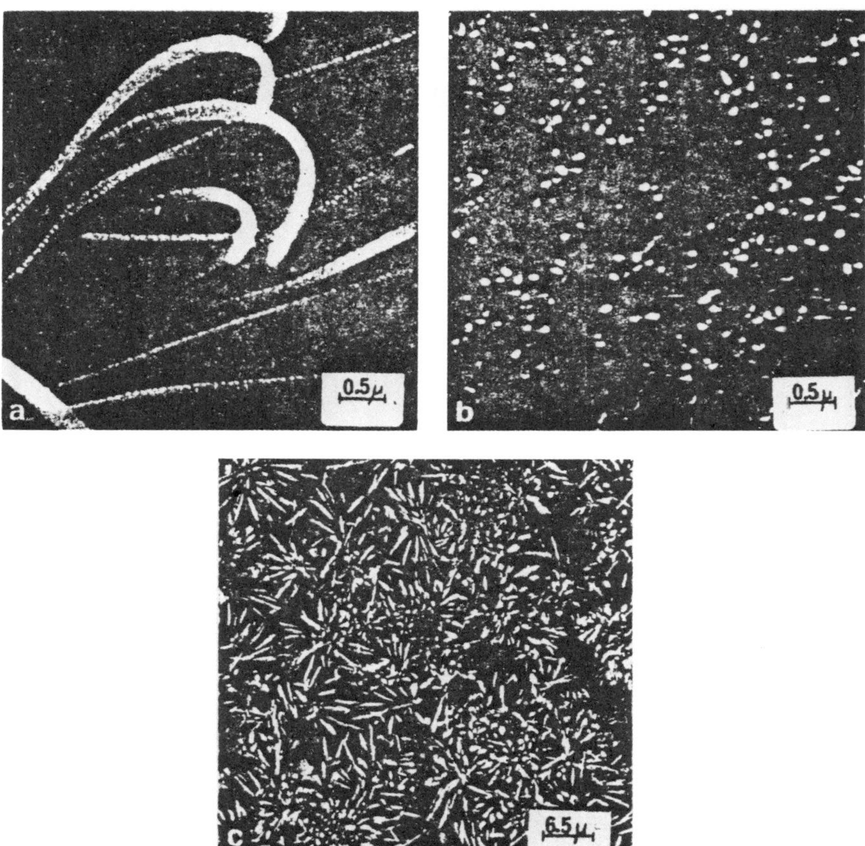


Fig. 13. SEM of the aged gels: (a) 15N for two days; (b) 15N aged for one month inside a dessicator; (c) 40N aged for two days.

that the regions where the crystallization occurs, are very limited and occur preferentially at the surface, which is in direct contact with air. This behaviour was confirmed by X-ray diffraction.

The samples 15N, 20N, 25N and 40N were exposed, on their supports, in air for one month. The X-ray diffraction spectra were identical to the latter spectra which are shown in figs. 9, 10, 11 and 12, respectively. On the other hand, the spectra of the similarly treated samples after the homogenization of the powders display considerable changes (fig. 14). It can be seen that there is an apparent decrease of the intensity of the diffraction peaks of the gels 20N, 25N and 40N. The gel 15N does not seem to be affected by the homogenization of the powder.

This variation of the spectra indicates that the "surface" layers of the gel in

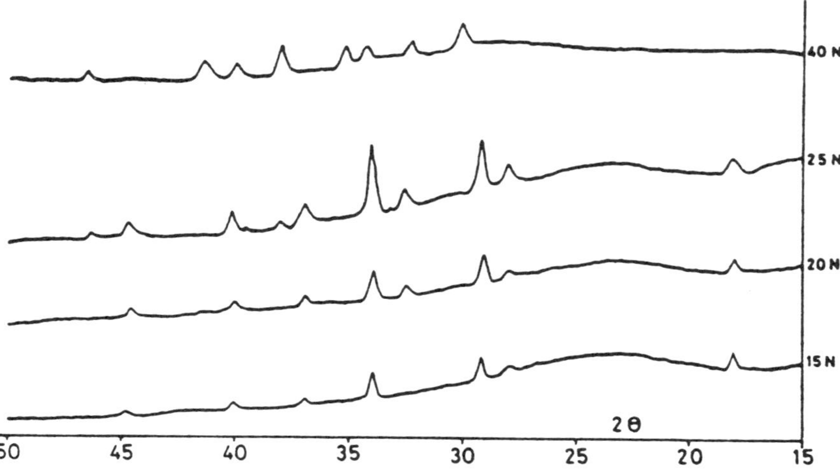


Fig. 14. X-ray diffraction spectra of the gels aged for one month after homogenization.

contact with air contain more carbonates than those situated in the interior. The homogenization of the powder produces, as a consequence, a redistribution of the crystals over the whole sample.

Fig. 15 shows the relative variation (before and after homogenization) in the intensity of the peak ($d = 2.63$ Å) for the samples 15N, 20N and 25N and that of the peak ($d = 2.76$ Å) for the gel 40N. This variation is proportional to the concentration of $Na_2O$. It is only significant above 15 mol% $Na_2O$.

It is not surprizing that it can be deduced that the concentration of the carbonate differs from the inside to the surface of the gels. It has often been indicated that, in the binary glasses $SiO_2-Na_2O$, the diffusion of the $Na^+$ ions takes place from the internal layer towards the surface. This is also true for an attack by an aqueous solution [24] as well as by atmospheric agents ($CO_2$ and $H_2O$) [25]. The migration of $Na^+$ ions must be more rapid in a gel than in a glass, because the gel has an open texture which favours this diffusion.

To examine the structural changes arising from the crystallization of Na-carbonates, IR reflection spectroscopy (IRRS) measurements were made.

The evolution of the IRRS spectra of the gels 15N, 20N, 25N and 40N during the ageing in air is shown in figs. 16 and 17.

The changes in the reflection bands located between 9 and 12 $\mu$ are altogether indicative of the corrosive attack of the gels by atmospheric agents (fig. 16). Indeed the (S) band increases in intensity during the first two days and at the same time shifts towards a higher wavenumber. This band is located at 1070 $cm^{-1}$ after five days exposure in air, for gel 15N as well as for gel 20N. It exactly corresponds to the position of the (S) band of the pure silica gel (fig. 7). An analogous behaviour, observed in the corrosion study of the $Na_2O2SiO_2$ glass using aqueous solutions, has been attributed to the disap-

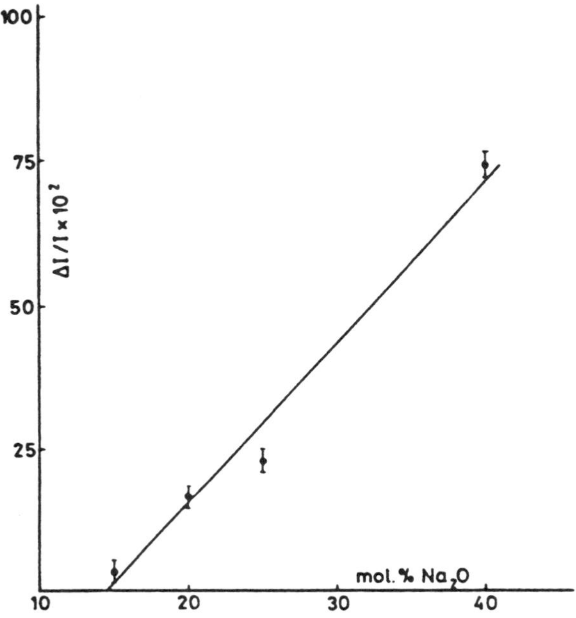


Fig. 15. Relative variation of the intensity of the diffraction due to carbonate before and after homogenization of the powder as a function of the $Na_2O$ content.

$$\left(\frac{\Delta I}{I} = \frac{I_{bef.} - I_{aft.}}{I_{bef.}} \times 10^2\right).$$

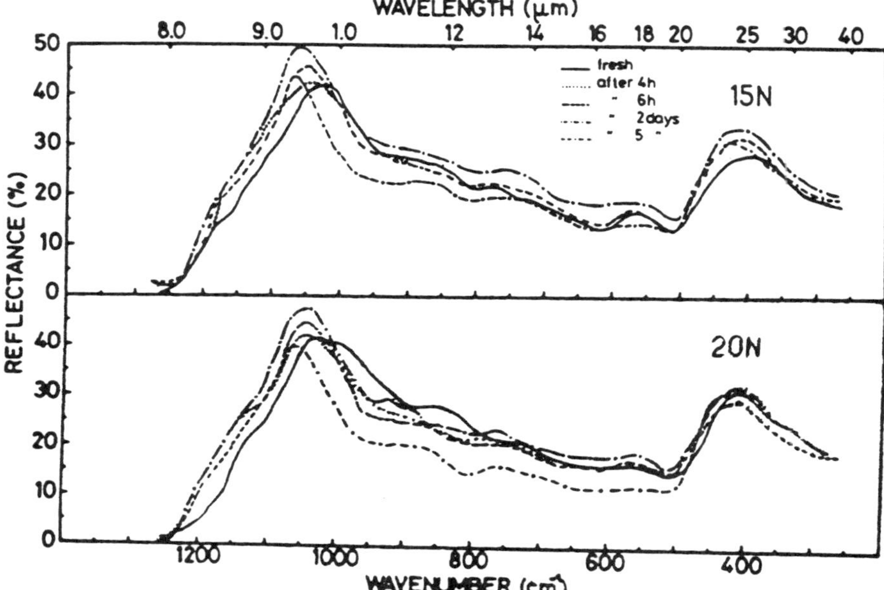


Fig. 16. IRRS of the gels 15N and 20N as a function of their ageing.

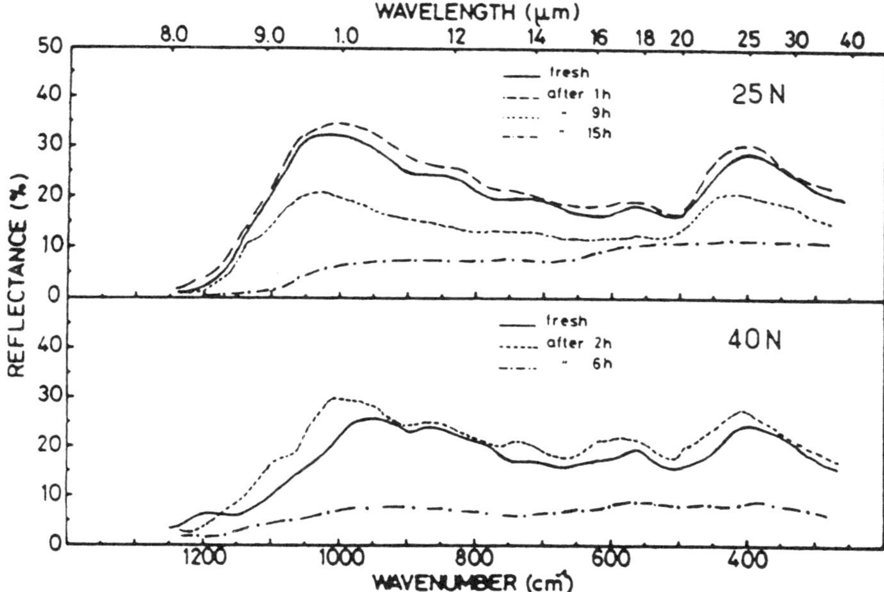


Fig. 17. IRRS of the 25N and 40N gels as a function of their ageing.

pearance of the $Na^+$ ions from the surface layers of the glass to the solution. The exchange with the $H_3O^+$ ensures the neutrality of the non-bridging oxygens ($Si-O^-$) [24,26].

The intensity of the (S) band decreases after five days of exposure in air in the samples 15N and 20N (fig. 16). This is probably due to the fact that the $Na^+$ ions remain at the surface of the gel in the form of carbonate which produces a loss of intensity by the light scattering.

For the same reasons, the spectra of gels 25N and 40N do not show any significant reflections after few hours (fig. 17).

## 4. Conclusion

A structural study of gels $x Na_2O-(1-x)SiO_2$ has shown that these materials are particularly unstable under ambient conditions.

These gels, which are initially amorphous, display a pronounced reactivity with respect to certain atmospheric agents, i.e, $CO_2$ and $H_2O$.

Their interaction with the environment gives rise to the crystallization of Na-carbonates. The nature of these crystallized carbonates is related to the concentration of $Na_2O$ in the gels. When the $Na_2O$ content is $\leq 25$ mol% the carbonate which precipitates is trona, while it is thermonatrite for compositions richer in $Na_2O$ after exposure to air for about one month.

The corrosive action of air is restricted to the surface layer of the material, except in the case of the gel 15N. This difference in behaviour can be explained by a diffusional process of the $Na^+$ ions from the internal layers towards those which are in contact with air.

The content of $Na^+$-ions in the internal layers decreases as they are replaced by $H^+$ ions. It is possible, however, that the physically adsorbed water plays a role in this transport.

The evolution of the IRRS during ageing in air is altogether analogous to that usually found in the corrosion studies of the binary glasses $Na_2O$–$SiO_2$ employing aqueous solutions. However, in the case of the gel having a large porosity, this evolution is extremely rapid.

A comparison of the structure and the behaviour in the presence of a corrosive medium between glasses and the gels of same compositions, shows that a great similarity exists between these materials.

## References

[1] J.D. Mackenzie, Modern aspects of the vitreous state, Vol. 3 (Butterworth, London, 1964) p. 149.
[2] R. Roy, J. Am. Ceram. Soc. 52 (1969) 344.
[3] H. Dislich, Angew. Chem. Internat. Edit. 10 (1971) 363.
[4] B.E. Yoldas, J. Mat. Sci. 12 (1977) 1203.
[5] K. Kamiya and S. Sakka, Res. Rep. Fac. Eng. Mie Univ. 2 (1977) 87.
[6] M. Yamane, S. Aso, S. Okano and T. Sakaino, J. Mat. Sci. 14 (1979) 607.
[7] S.P. Mukherjee, J. Non-Crystalline Solids 42 (1980) 477.
[8] M. Decottignies, J. Phalippou and J. Zarzycki, J. Mat. Sci. 13 (1978) 2605.
[9] R. Jabra, J. Phalippou and J. Zarzycki, J. Non-Crystalline Solids 42 (1980) 403.
[10] R. Jabra, J. Phalippou and J. Zarzycki, Rev. Chim. Min. 16 (1979) 245.
[11] J. Phalippou, M. Prassas and J. Zarzycki, Verres et Refract. (Nov. 1981).
[12] R.K. Iler, The chemistry of silica (Wiley-Interscience, New York, 1979) pp. 281, 366.
[13] E.R. Lippencott, A. Van Valkenburg, C.E. Weir and E.N. Bunting, J. Res. Nat. Bur. Stand. 61 (1958) 61.
[14] R. Hanna, J. Am. Ceram. Soc. 48 (1965) 595.
[15] F. Matossi, J. Chem. Phys. 17 (1949) 679.
[16] J. Fripiat and A. Jelli, Congrès Intern. du verre, Bruxelles (1968) p. 14.
[17] R. Hanna and G.J. Su, J. Am. Ceram. Soc. 47 (1964) 597.
[18] J. Zarzycki and F. Naudin, J. Chim. Phys. 58 (1961) 830.
[19] J.R. Ferraro and M.H. Manghnani, J. Appl. Phys. 43 (1972) 4595.
[20] J.R. Sweet and W.B. While, Phys. Chem. Glasses 10 (1969) 246.
[21] J. Fraissard and B. Imelik, J. Chim. Phys. (1962) 415.
[22] L.L. Hench, M. Prassas and J. Phalippou, J. Non-Crystalline Solids, submitted for publication.
[23] Ref. (12), p. 117.
[24] L.L. Hench and D.E. Clark, J. Non-Crystalline Solids 28 (1977) 83.
[25] H. Cohen and J.W. Park, Phys. Chem. Glasses 22 (1981) 39.
[26] R. Doremus, J. Non-Crystalline Solids 41 (1980) 145.

# PREPARATION OF $xNa_2O$-(1 - x)$SiO_2$ GELS FOR THE GEL-GLASS PROCESS: II. THE GEL-GLASS CONVERSION

M. Prassas* and J. Phalippou

*Laboratory of Materials Science and CNRS Glass Laboratory*
*University of Montpellier II, France*

L.L. Hench

*Ceramics Division, Department of Materials Science and Engineering*
*University of Florida*
*Gainesville, FL 32611*

Abstract

$SiO_2$ gels containing from 317 to 40 mole % $Na_2O$ have been synthesized by the hydrolysis and polycondensation of metal organic precursors. The effect of $Na_2O$ content on thermal processing variables includes: evacuation of residual organic residues, decomposition of carbonates ($T_{ce}$), beginning of densification ($T_{db}$) and initiation of crystallization ($T_{ci}$). Thermal control of the gel-glass conversion is shown to be a function of $Na_2O$ content and $\Delta T_v = T_{ci} - T_{db}$. At high $Na_2O$ content (>33 mole %) $\Delta T_v$ is very small whereas at the $Na_2O$-$SiO_2$ eutectic composition it is nearly 200°C. By using this method $Na_2O$-$SiO_2$ glasses are obtained in the range of 440°-640°C.

*Currently at Corning Europe Inc. Centre Europeen de Recherche, B.P.3, 77211 Avon Cedex, France

## I. Introduction

Practical aspects of the preparation of dense glasses via a gel route has attracted attention by several investigators. As a result both single oxide and multicomponent oxide gels have been successfully prepared and converted to dense amorphous materials by various procedures.[1-3] Special importance has been given recently to the preparation of glass by sintering of monolithic gels.[4-6] This technique is recognized to be advantageous as a means for making very pure refractory materials at low process temperatures. For example, high purity vitreous silica can be synthesized at temperatures as low as 1150-1250°C.[6]

In addition to its potential technological value the "gel-glass" conversion raises important scientific concerns about the structure of glass and the nature of glass formation. In fact, the synthesis of glass by melting crystals and subsequent cooling of the supercooled liquid versus densification of gels to obtain "glass-like" materials, are two diametrically opposed processes. In the first case, the structural amorphous state is retained from high temperatures. In the second case vitreous network formation occurs at room temperature by chemical polymerisation.

The gel-glass conversion is achieved at temperatures slightly higher than the $T_g$ of the corresponding melted glass (Fig. 1), probably through an intermediate supercooled liquid phase. The successful formation of glass by sintering of gels is the result of a competition between phenomena which lead to densification and those which promote crystallization. It has been shown that both the densification process and the kinetics of crystallization depend essentially on the time-viscosity ratio.[7] Nevertheless in practice, other phenomena inherent to the physico-chemical nature of the starting gels appear during the densification heat treatment and can play an important role in the "gel-glass" conversion.

For example, the complete elimination of the residual liquid inside the pores and the oxidation of the organic groups affect the integrity of the initial material. The -OH content, and alkaline modifiers alter the viscosity of the gel and consequently affect the sintering and crystallization process. In addition, the experimental procedure followed during preparation of gels is now recognized to have a significant effect on the "gel-glass" conversion.[8] Each of these factors influence the physico-chemical properties of the initial gelled material and therefore play an important role in its conversion to a dense glass.

The purpose of the present investigation is to study the "gel-glass" conversion in relation to the chemical composition of the binary system $Na_2O$-$SiO_2$. A wide range of compositions from 3.7 mole % at 40 mole % $Na_2O$ have been prepared previously and their structural characterization presented in a previous paper.[9]

## II. Experimental Procedure

### II.1 Gel preparation

Tetramethoxysilane and sodium methylate were used as precursors of $SiO_2$ and $Na_2O$ oxides. Hydrolysis and polycondensation of the mixtures were conducted at 0°C. The resulting monolithic gels, after ageing and drying, were kept in a desiccator under a low vacuum. Details of the preparation procedure have been given elsewhere.[9] In the present study samples will be designated by xN with x = molar percentage of $Na_2O$; e.g. a 25N gel or 25N glass refers to 25 mol % $Na_2O$–75% $SiO_2$.

### II.2 Characterization techniques

Structural evolution of the gels as a function of temperature was followed by x-ray diffraction. Crystalline species were identified by comparison with the ASTM standard data files. A differential electronic dilatometer (ADAMEL DI.10) was used to measure linear dimensional changes of the solids to monitor the sintering of the gels. The measurements were made with a heating rate of 3°C $min^{-1}$. Parallelepiped samples with 10-15 mm length were used.

Additional information concerning the chemical and structural changes during the thermal treatment of the gels was obtained using DTA. Calcined $Al_2O_3$ was the reference material. The heating rate was 15°C $min^{-1}$.

## III. Results and Discussion

### III.1 Chemical and structural changes

During thermal treatment of gels, several phenomena occur. Dehydratation, oxidation of the residual organic groups, carbonate decomposition, sintering and crystallization are the most important. In order to determine the temperature range at which those chemical and structural changes appear, DTA and x-ray diffraction measurements were undertaken.

#### III. 1.1 Differential Thermal Analysis (DTA)

The thermograms of the gels are generally not easily interpreted. They frequently show significant drift of the base line and assignment of small intensity thermal effects is difficult. Figure 2 shows the DTA thermograms of the 15N, 20N, 25N, 33N and 40N gels.

The following thermal events can be identified:

a) For all samples, a large endothermic effect between 90 and 100°C appears systematically. It disappears during a second DTA measurement. This effect, which is more intense for aged gels, is due to the evaporation of the water-methanol liquid located inside the porous solid.

b) Between 230° and 250°C an exothermic effect appears for the gels 15N, 20N and 25N. Its position is shifted to 300°C in the case of 33N and 40N samples. At the same temperature range, thermograms of pure silica gels also show a similar exothermic effect which was related to the following oxidation reaction.[10]

$$\equiv Si\ OCH_3 \xrightarrow{O_2} \equiv SiOH \qquad (1)$$

c) At higher temperatures crystallization occurs. It appears in the thermograms as an exothermic effect at 509°, 520° and 605°C for the 40N, 33N and 25N gels respectively. This effect is less intense for the gels with lower concentrations of $Na_2O$. During the cooling of samples with higher silica content, the $\alpha \rightleftharpoons \beta$ inversion of cristobalite appears around 210°C on the DTA curves.

d) Endothermic effects between 730 and 830°C are related to the melting of the crystals precipitated during the thermal treatment.

e) Decomposition of sodium carbonate crystallities, identified by x-ray diffraction, within the porous gel structure also occurs during a thermal cycle. The temperature of this decomposition does not show up clearly on the DTA curves probably because it is not a true decomposition reaction such as occurs for a pure material. It is more likely to be a mixed reaction between $Na_2CO_3$ and the $SiO_2$ network and therefore takes place over a wide temperature range. Thus the rate of the reaction, and the heat evolution, depends strongly on gel parameters such as porosity and specific area.

### III. 1.2 X-ray diffraction results

a) Structural evolution for $25 < T < 500$°C

The x-ray diffraction spectra of all the compositions studied show evidence of $Na_2CO_3$ and $NaHCO_3$ crystals from 85°C to 300°C. The structural evolution of the 15N and 40N gels given in the Figure 3 is typical of all the compositions. The samples were heated at 2.5°C $min^{-1}$ and kept 1 hour at the given temperature.

Figure 3 shows that beyond a certain temperature, 400°C for 15N and 450°C for 40N, the spectra is a diffuse pattern characteristic of an amorphous state. The temperature at which the x-ray spectra of the gels no longer show any diffraction peak due to the Na-carbonates, is a characteristic processing parameter and will be designated by $T_{ce}$. The value of $T_{ce}$ depends on the chemical composition of the gel. If we assume that the following reaction occurs during thermal treatment, the complete elimination of the $Na_2CO_3$ by reaction with $SiO_2$ (eq. 2)

$$(Na_2CO_3 + SiO_2 \rightleftharpoons \text{Na-Silicate} + CO_2\uparrow) \qquad (2)$$

should also be related to the degree of ease in escape of the $CO_2$. Consequently, the 25N gel which is the most porous (75%) eliminates the carbonates at temperatures as low as 250°C (Table 1). Gels with a large concentration of sodium carbonates tend to have a higher $T_{ce}$ because there are more carbonate species to decompose.[9]

b) Structural evolution at T> 500°C.

Another important parameter to control in the gel-glass conversion is the temperature at which crystallization occurs. Since crystallization is a time-temperature dependent process, the temperature for initiation of crystallization ($T_{ci}$) is determined in 1 hour treatments in a manner similar to the $T_{ce}$ determination. To avoid any superposition of phenomena which could affect the crystallization kinetics, the gels were preheated at 400°C for two hours. The values of $T_{ci}$ versus concentration of $Na_2O$ are given in Table 1. The first value corresponds to completely amorphous gels whereas the second indicates the temperature at which a small intensity diffraction peak appears in the x-ray spectra.

We have seen in previous studies that a minor addition of $Na^+$ ions (4000 ppm) greatly decreased the devitrification temperature of a silica gel.[11] In this case the gel-glass conversion could not be achieved due to the onset of devitrification.[11] However, the results of Table 1 show that there is little effect of increasing amounts of $Na_2O$ on the initiation of crystallization until as much as 33 mole % is present.

At temperatures higher than $T_{ci}$, the gels crystallize and gel-derived ceramics can be obtained. However at temperatures close to the corresponding liquidus, the crystallization rate is low, so the gel-glass conversion can be conducted successfully by a rapid sintering and cooling schedule. Figure 4 shows a typical structural evolution of the 20N sample between 640 and 1050°C.

Table 1

| Gel | Porosity (%) | $T_{ce}$(°C) | $T_{ci}$(°C) | $T_{db}$(°C) |
|---|---|---|---|---|
| 3.7N | ..--- | ..--- | 640-660 | 610 |
| 7.7N | --- | --- | 620-640 | 545 |
| 12.7N | --- | --- | 620-640 | 510 |
| 15 N | 48 | 280-300 | 620-640 | 455 |
| 20 N | 68 | 280-300 | 600-620 | 460 |
| 25 N | 75 | 240-260 | 620-640 | 450 |
| 33 N | 61 | 450-480 | 500-525 | 440 |
| 40 N | 66 | 400-420 | 440-460 | 420 |

At 640°C cristobalite is the primary crystalline phase. An increase in the rate of crystallization at 700°C is indicated by the substantially increased intensity of the x-ray diffraction peaks. At 700°C a sodium disilicate crystal phase is identified. However, this phase disappears at higher temperatures. For temperature higher than 800°C we observe a decrease in the cristobalite peaks, and a concomitant appearance and growth of tridymite from 885°C upwards. Both allotropic phases of silica are developed until 1050°C, a temperature where the crystallization rate is low.

Finally, as expected the material remains amorphous if it is treated directly at temperatures slightly higher than 1050°C. The behavior described for the 20N gel (Fig. 4) is quite similar to that of other gel compositions. However, the nature of the crystals precipitated depends on the $Na_2O$ concentration in the gel (Figure 5). The value of $T_{ci}$ for gels with < 25 mole % $Na_2O$ appears controlled by the $SiO_2$-$Na_2Si_2O_5$ eutectic temperature whereas the higher $Na_2O$ gels show a progressively decreasing $T_{ci}$. The phases developed upon crystallization are as predicted from the phase equilibria in the $Na_2O$-$SiO_2$ system[15] except cristobalite and tridymite occur at much lower temperatures than under equilibrium conditions.

III. 1.3 Dilatometric behavior

In a previous paper, the detailed study of the gel-derived 33N composition, we saw that densification starts around 420-450°C.[12] The dialotometric shrinkage curve of this composition shows at the same range of temperatures an abrupt change of the shrinkage rate (Figure 6). A few degrees above 420-450°C the gel can be converted to a "glass" very rapidly, whereas only a small change in the density is observed at 20-30°C below this temperature even for long-term thermal treatments.

We have calculated using the shrinkage data, the viscosity of this characteristic gel-glass transformation temperature.[13] Its value is similar to the viscosity associated at the vitreous transition range ($10^4$-$10^{12}$Poise)[13] Thus this temperature, which corresponds to the beginning of the densification process by a viscous flow mechanism, can be estimated easily by the shrinkage data. We will call it $T_{db}$.

Because $T_{db}$ is related to a given viscosity range, we expect that its value will be a function of the $Na_2O$ concentration of the gel. In fact as shown in Figure 7 which summarizes the shrinkage curves of all the compositions, $T_{db}$ decreases from 610°C to 420°C with the increase of the $Na_2O$ concentration from 3.7 to 40 mole %. It is interesting to note here that the $T_{db}$ value falls more rapidly in the concentration range 3.7-15 mole % of $Na_2O$ whereas it seems to be relatively constant for the higher $Na_2O$ concentrations (Figure 8).

Although the part of the shrinkage curve above $T_{db}$ indicates sintering by a viscous flow mechanism, the first portion of the curve between 25°C and $T_{db}$ does not seem to have any apparent dependence on the gel composition. In addition we observed that this part of the shrinkage curve is not reproducible unless a given gel composition has the same processing "history". This fact is most pronounced for the gels with a high concentration of $Na_2O$.

As an example of the effect of processing history, Figure 9 shows the shrinkage curves of a 40N sample which has undergone ageing in an ambient atmosphere for three different times, non-aged, 5 hrs, and 50 hours. For the non-aged gel the shrinkage begins around 100°C and its rate increases slightly with increasing temperatures until 420°C. On the other hand, the gel with limited aging, (5 hours in air) shows a large shrinkage (6-7%) from 100 to 200°C. Shrinkage occurs even more rapidly for the sample aged for 50 hours.

This behavior can be understood if we assume that in the $Na_2O$-$SiO_2$ system the gel network has considerable mobility due to the large number of non-bridging oxygen ions and deficient Si-O-Si bonds. Consequently the gels swell easily by $H_2O$ and $CO_2$ sorption on Si-O-Na sites and shrink easily during desorption of gases. Because of this apparent "plasticity", gel samples with large $Na_2O$ content do not break during shrinkage.

Removal of liquid from the pores of a 40N sample by sucessive heating and the loss of sorption capacity due to partial sintering causes the first part of the shrinkage curve to disappear (Figure 10). The reaction of Na-carbonates, achieved during the thermal treatment, as we have seen previously by x-ray diffraction, seems to be responsible for the little shrinkage or even small apparent expansion observed between 200°C and $T_{db}$ (Figures 5, 9 and 10). In fact, the Na-carbonate crystals cover the surface of the particles[9] and do not allow the thermally induced condensation between neighboring hydroxyl groups. At temperatures higher than 300°C, the reaction between the Na-carbonate crystals and the silica network occurs. The gaseous products of this reaction and the water resulting from thermal condensation of -OH groups at this temperature range cannot be eliminated easily (especially when the heating rate is too high) and a relative expansion of the material is observed. Residual gaseous species are the most difficult to evolve, because of partial pore closure during previous runs, Figure 10, and consequently more expansion is observed prior to final shrinkage.

These results are valid for gels of more than 10 mole % of $Na_2O$ and illustrate the importance of controlled heating and atmospheric schedules to produce a fully dense alkali-silicate glass from gel processing.

III. 1.4 The gel-glass conversion

The above results provide the basis for establishing an optimum schedule in order to densify the gel into glassy materials. We have defined previously three characteristic temperatures which are important processing parameters:

1) The temperature at which Na-carbonates are eliminated ($T_{ce}$)
2) The temperature for initiation of crystallization ($T_{ci}$)
3) The temperature at which densification begins ($T_{db}$)

These parameters can be combined to produce a thermal processing diagram. Such a processing temperature-$Na_2O$ concentration diagram shows four regions whose boundaries are curves describing $T_{ce}$, $T_{ci}$ and $T_{db}$ as a function of mole % $Na_2O$ (Figure 11). The gel-glass conversion can be achieved by sintering only in the region between the $T_{db}$ and $T_{ci}$ curves. Figure 11 shows that extrapolation of the $T_{db}$ and $T_{ci}$ curves indicates that gel-glass conversion for gels of less than 2 mole % of $Na_2O$ is not possible at low temperatures. In fact in this case the temperature at which the gel has an appropriate viscosity to be densified ($T > T_{db}$), is below that at which crystallization occurs. For the higher concentrations (>30 mole % of $Na_2O$) sintering begins when the Na-carbonates have not been completely eliminated. Therefore a preheating of the gel for a long time at $T < T_{db}$ should be applied in order to obtain a homogeneous "glass".

Based upon the above analysis, a two-step densification schedule is adopted:

1) Preheat the samples at 2.5°C $min^{-1}$ until a temperature $T = T_{db}-30$ for 100 hours.

2) Increase the temperature rapidly (10°C $min^{-1}$) until $T_{db}$ and keep the gels at the densification temperature for 1 hour.

The gel-glass conversion is considered achieved when the density reaches a value which is 99% of the density of a glass prepared via melting. This densification schedule results in the conversion of the 15N, 20N, 25N and 33N gels to "glass-like" materials. The minimum densification temperature ($T_{db}$) is given in Table 1.

The physical properties (refractive index, density) of the glass derived gels are similar to those of the glass prepared by conventional methods. However 33N gel-derived glasses (as we should expect from the Figure 11) may retain a black coloration, a result of incomplete elimination of the Na-carbonates. The ability of the gels to be converted to glass depends on the range between the crystallization temperature and the densification temperature. Therefore, we can define a semi-empirical criterion for gel derived vitrification as the difference between the two temperatures; i.e., $\Delta T_v = T_{ci}-T_{db}$. The higher $\Delta T_v$, the easier the gel-glass conversion. As we can see from Figure 11 the eutectic composition has the higher value of $\Delta T_v$. According to our criteria this composition has the higher "vitrification ability" for all the compositions of the $Na_2O$-$SiO_2$ systems.

## Conclusions

A thermal processing diagram can be constructed for the $Na_2O$-$SiO_2$ gel -glass system. The diagram is based upon three characteristic temperatures:

1) $T_{ce}$, the temperature at which Na$^-$carbonates are eliminated; 2) $T_{ci}$, the temperature for initiation of crystallization; and 3) $T_{db}$, the temperature at which densification begins. Because $T_{db}$ is nearly compositionally invariant over a wide range from 15-35 mol % $Na_2O$ there is a broad compositional region where gel-derived glasses can be prepared at temperatures in the range of 440-640°C. By control of each of the three characteristic temperatures it is possible to produce glasses with a wide range of physical properties, due to variations in $Na_2O$ contents with the same low temperature processing cycles.

Acknowledgements

The authors kindly acknowledge the encouragement of Professor J. Zarzycki throughout the course of this work and two of them (L. L. Hench and M. Prassas) acknowledge financial support of U.S. AFOSR Contract #F49620-80-C-0047.

## References

1. H. Dislich, Angew. Chem. Internat. Edit. 10 (1971) 363.

2. B. E. Yoldas, J. Mat. Sci. 12 (1977) 1203.

3. K. Kamiya and S. Sakka, Res. Rep. Fac. Eng. Mie Univ. 2 (1977) 87.

4. M. Yamane, S. Aso, S. Okano and T. Sakaino, J. Mat Sci. 14 (1979) 607.

5. L. Klein and G. Garvey, J. NonCryst. Solids 48[1982] 96-104.

6. J. Zarzycki, M. Prassas and J. Phalippou, J. of Mater. Sci. 17(1982) 3371-3379.

7. J. Zarzycki, J. Non-Cryst. Solids 48 (1982) 105.

8. B. E. Yoldas, J. Amer. Ceram. Soc. 65 (1982) 387.

9. M. Prassas, J. Phalippou, L. L. Hench and J. Zarzycki, J. Non-Cryst. Solids 48 (1982) 79.

10. J. Phalippou, M. Prassas and J. Zarzycki, Verres et Refract. 35 (1981) 975.

11. J. Phalippou, M. Prassas and J. Zarzycki, J. Non-Cryst. Solids 48 (1982) 17.

12. L. L. Hench, M. Prassas and J. Phalippou, J. Non-Cryst. Solids in press).

13. M. Prassas, Unpublished work.

14. M. Decottignies, J. Phalippou and J. Zarzycki, J. Mat. Sci. 13 (1978) 2605.

15. in E. M. Levin, C. R. Robins and H. F. McMurdie, Phase Diagrams for Ceramists, The American Ceramic Society, Columbus, OH, 1964.

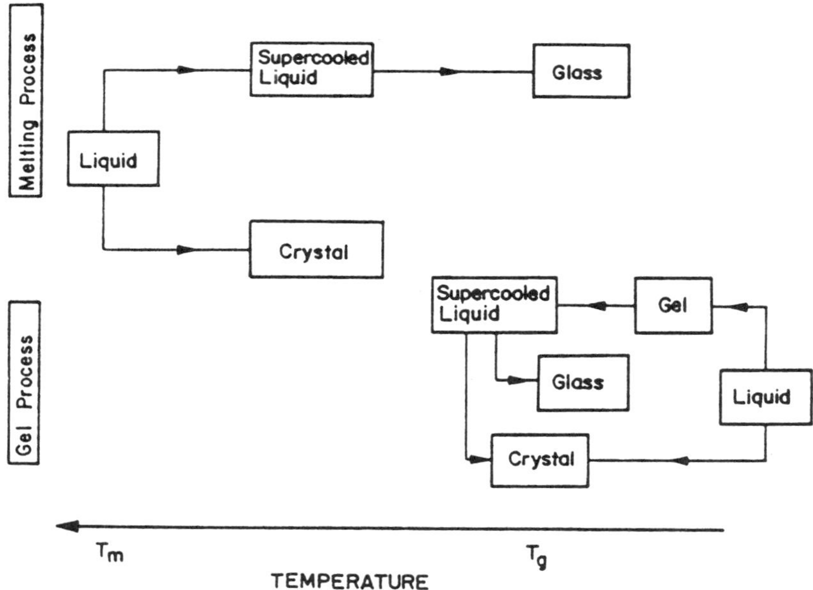


Fig. 1. Schematic representation of the gel and the melting process to obtain glass.

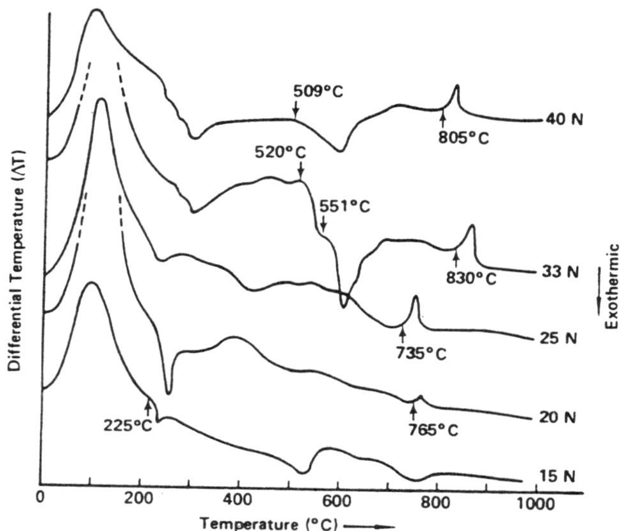


Fig. 2. DTA thermograms.

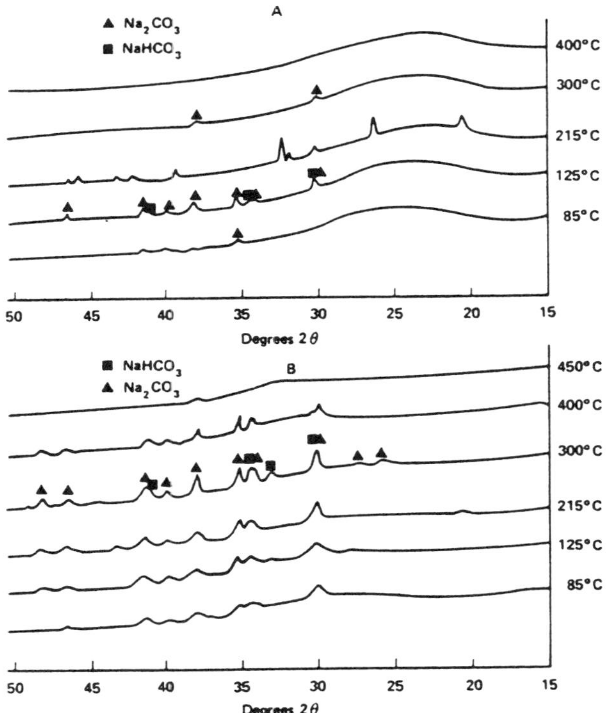


Fig. 3. X-ray diffraction spectra as function of the temperature of densification.

A: Gel 15N
B: Gel 40N

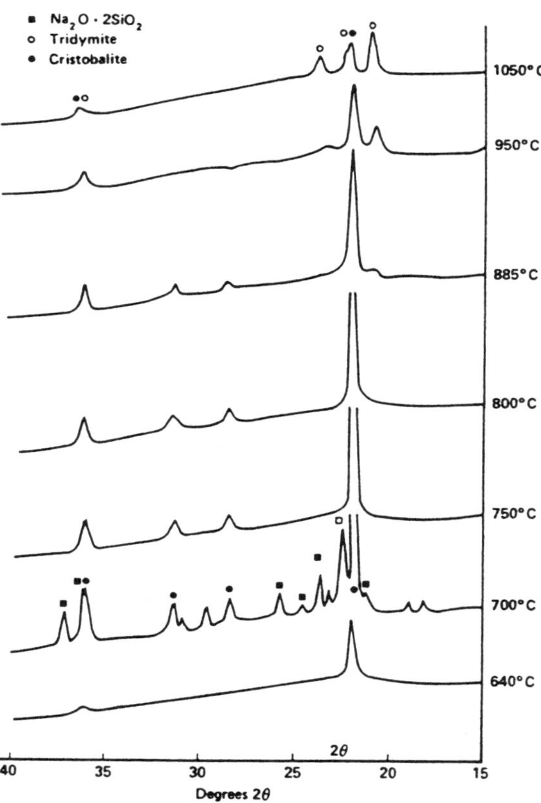


Fig. 4. X-ray diffraction spectra for the 20N gel as function of the temperature of densification.

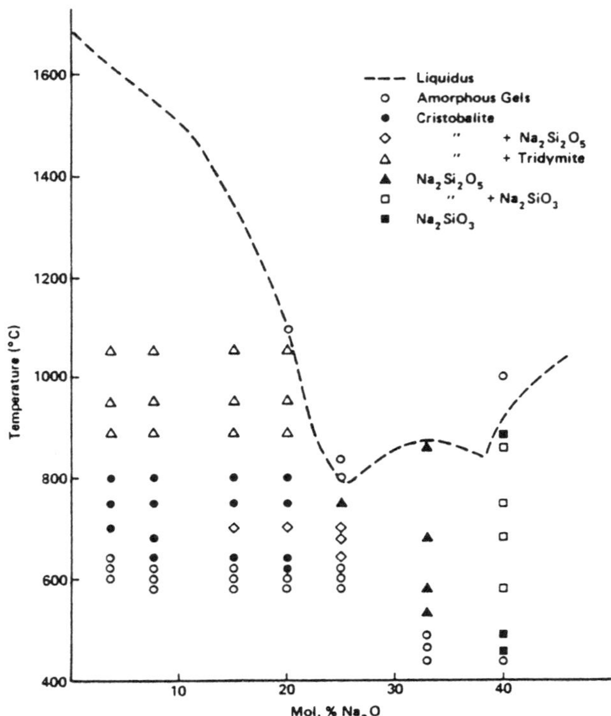


Fig. 5. Crystalline phases of the gels $xNa_2O-(1-x)SiO_2$ as function of the temperature (1 hour treatment) of densification.

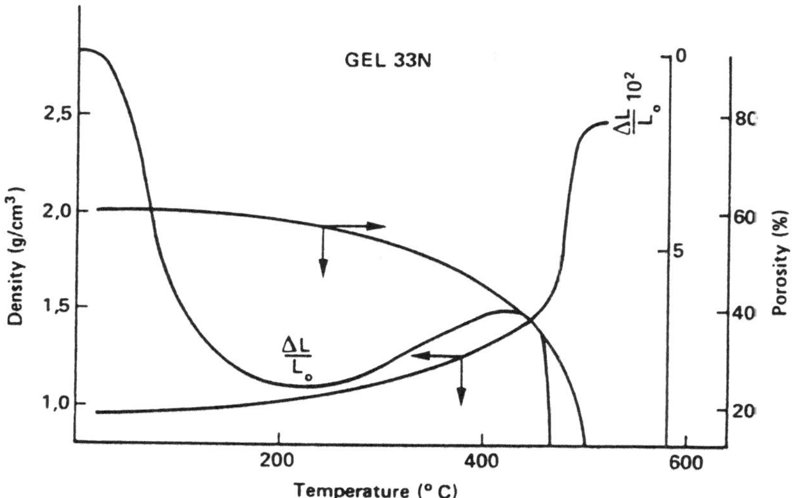


Fig. 6. Density, porosity (after {12}) and shrinkage for the 33N as function of the temperature of densification.

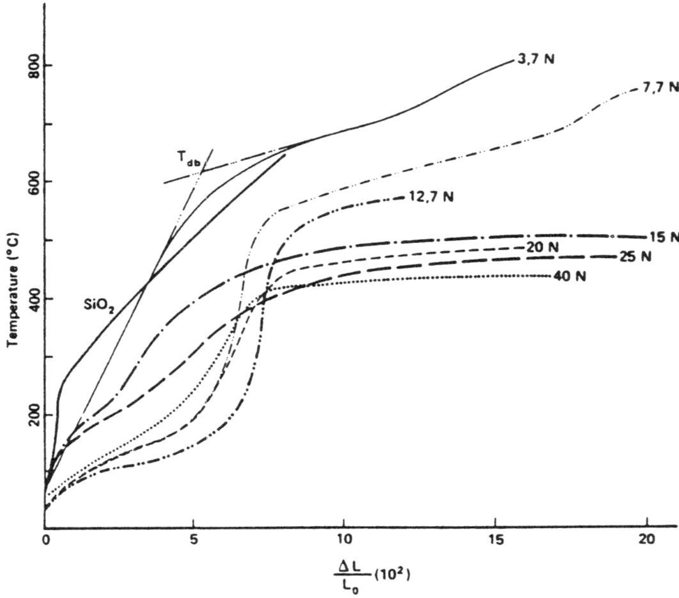


Fig. 7. Dilatometric shrinkage curves for $Na_2O-SiO_2$ gels.

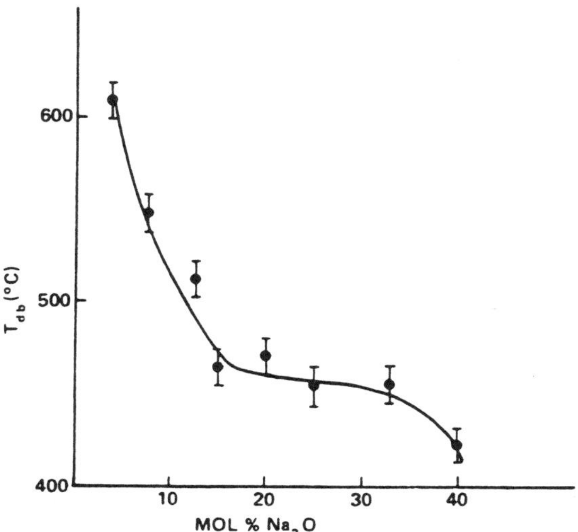


Fig. 8. Temperature of the densification beginning ($T_{db}$) as function of the $Na_2O$ concentration.

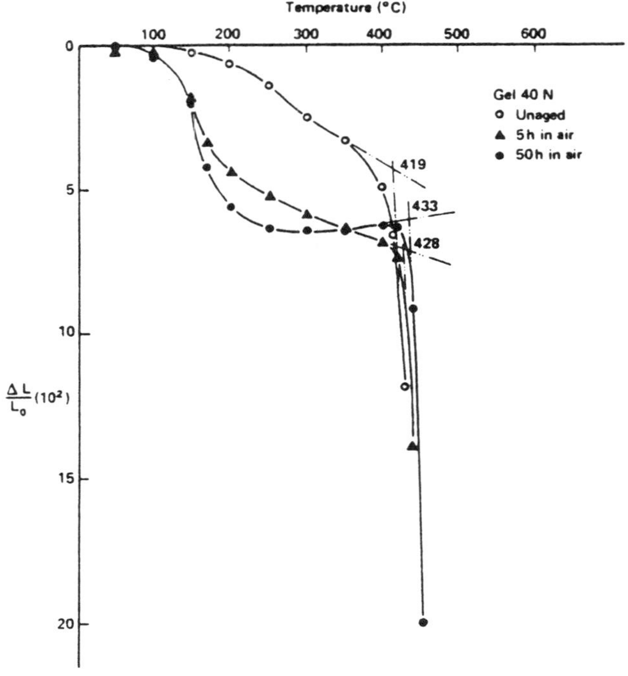


Fig. 9. Dilatometric shrinkage curves for gel 40N with and without aging.

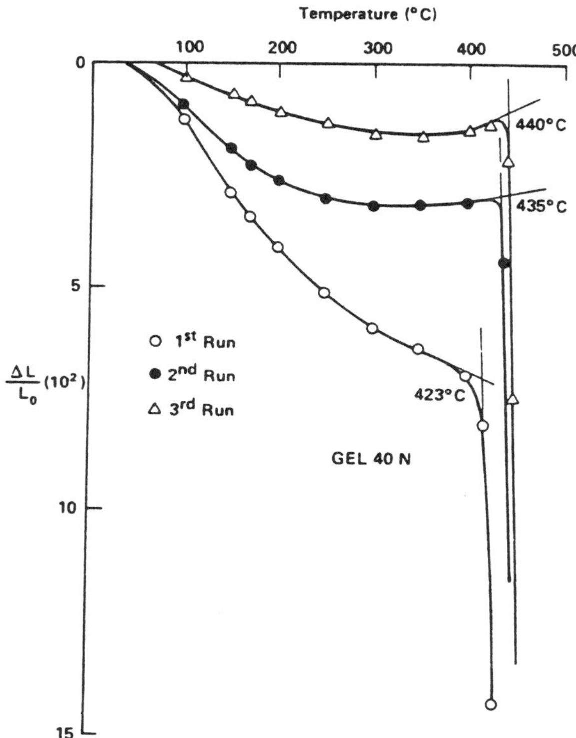


Fig. 10. Effect of repeated dilatometric shrinkage runs on gel 40N densification temperature.

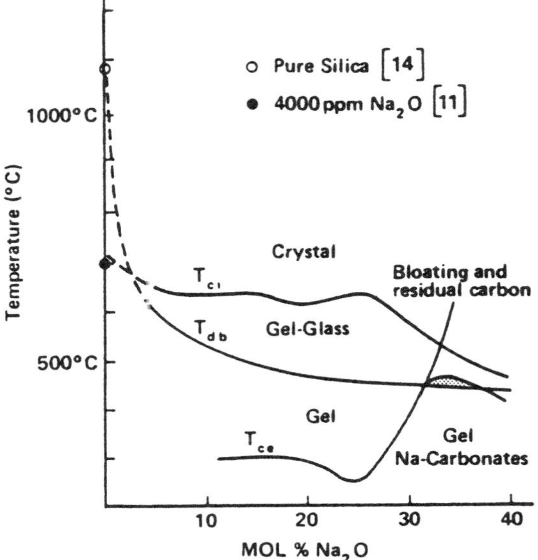


Fig. 11. Gel-glass processing diagram.

# PROTONIC CONDUCTION IN ALKALINE EARTH METAPHOSPHATE GLASSES CONTAINING WATER

**Yoshihiro Abe and Hajime Shimakawa**

*Department of Inorganic Materials, Nagoya Institute of Technology*
*Gokiso-cho, Showa-ku, Nagoya 466, Japan*

**L.L. Hench**

*Department of Materials Science and Engineering*
*University of Florida*
*Gainesville, FL 32611*

D.c. electrical conductivity data were obtained for $MO \cdot P_2O_5$ glasses ($M$ = Be, Mg, Ca, Sr, Ba) containing small amounts of water. The results suggest that the mobility of protons in the glasses increases with decreasing O–H bonding strength. The relation between the proton concentration and the conductivity or the apparent activation energy was studied for calcium metaphosphate glasses containing various amounts of water. The conductivity was found to be proportional to the square of the proton concentration; the apparent activation energy decreased linearly with increasing logarithm of proton concentration.

## 1. Introduction

Electrical charge carriers of oxide glasses such as $V_2O_5$–$P_2O_5$ or FeO–$P_2O_5$ are usually electrons but in glasses such as $Li_2O$–$SiO_2$ and $Na_2O$–CaO–$SiO_2$ charge carriers are usually mobile cations [1]. However, electrical conductivity measurements are also often affected by the presence of –OH groups, as for example, in lead silicate [2], calcium silicate [3], barium aluminium borate [4] glasses. Namikawa and Asahara [5] suggested that the electrical carriers in barium phosphate glasses are protons but as yet a quantitative relationship between conductivity and proton concentration has not exactly been shown. No systematic work concerned with protonic conduction in glasses has been done to date. It is possible that the conductivity by protons depends not only on the concentration but also on the strength of O–H bonding or OH...O hydrogen bonding.

In the present paper, dc electrical conductivity data are obtained for $MO \cdot P_2O_5$ glasses ($M$ = Be, Mg, Ca, Sr, Ba) containing water but not alkali metal ions. The relationships between the proton concentration and the conductivity or the apparent activation energy are studied for calcium metaphosphate glasses containing various amounts of water, where the peak wavenumber of IR absorption band due to O–H stretching is essentially the same ($\nu = 2920\ cm^{-1}$). It is known that peak wavenumber ($\nu$) of IR absorption bands due to O–H stretching is a measure of O–H bonding strength [6]. Further, a relationship between the mobility of protons and the peak wavenumber due to the OH bond ($\nu$) is also discussed for $MO \cdot P_2O_5$ glasses which have a different $\nu$ which depends on the kind of $M^{2+}$ ions (Be > Mg > Ca > Sr > Ba) [7]. The findings obtained may be useful in developing proton conductive glasses as a solid electrolyte separator for $H_2$–$O_2$ fuel cells.

## 2. Experimental

### *2.1. Glass preparation and dc conductivity measurement*

The glasses of composition $CaO \cdot P_2O_5$ (mol ratio) with variable OH content were prepared by melting reagent grade $Ca(H_2PO_4)_2 \cdot H_2O$ in a platinum crucible at 1200°C to 1300°C for various times (< 1 to 10 h) to control the state of hydration. The other metaphosphate glasses were obtained by melting a mixture of reagent grade metal carbonates and $H_3PO_4$ in a platinum crucible at 1100°C (Ba) to 1450°C (Be) for an hour. After melting, the glasses were cast in air onto a graphite slab and annealed in air at temperatures 20°C higher than the respective glass transition temperature for 20 min and cooled with the furnace. Discs (about 4 cm diameter × 0.4 cm thick) were cut from the blocks, ground, and polished with < 1 μm $Al_2O_3$ paste. Gold electrodes with double guard rings were evaporated onto the samples.

DC conductivity measurements were made with a vibrating reed electrometer (Takeda TR-43, impedance $10^{16}$) over a temperature range from 80°C to 200°C. The time required for steady state conduction depends on the conductivity and temperature of the sample. Usually, it was established within 30 min at 100°C and within 1 min at 200°C.

### *2.2. Determination of water content in glass*

Water content in the glasses was determined by infrared spectroscopy [8]. Plate specimens (about 1 cm × 1 cm × 0.05 cm thick) were prepared by cutting and polishing. Water content was estimated by the following Lambert Beer's Law

$$E = \epsilon c d \tag{1}$$

where $E$ is the absorbance ($= -\log T$, where $T$ = transmission); $\epsilon$ is the molar absorption coefficient ($l\ mol^{-1}\ cm^{-1}$); $c$ is the concentration (as $H_2O$, mol/l); and $d$ is the thickness of the specimen (cm). $E$ was obtained experimentally by using relation (2) to compensate for errors due to the surface roughness of the glass specimen,

$$E_\nu = \log(T_{3800}/T_\nu), \tag{2}$$

where $\nu$ is a given wavenumber. The value of $\epsilon$ depends on $\nu$ [6,8]; $\epsilon$ (as $H_2O$) for Be-, Mg-, Ca-, and Ba-metaphosphate glasses was assumed to be 70, 100, 120, and 130 l $mol^{-1}$ $cm^{-1}$, respectively.

## 3. Results and discussion

Fig. 1 shows IR spectra of calcium metaphosphate glasses, normalized to 100% transmittance at 4000 $cm^{-1}$ for thickness and loss according to eqs. (1) and (2), respectively. The broad bands at 2920 $cm^{-1}$ are due to O–H stretching. Scholze [6] established that there are three main bands due to OH or "water" in glass, i.e., Band 1 in the range 3640–3390 $cm^{-1}$, Band 2 (3000–2600 $cm^{-1}$), and Band 3 (2350 $cm^{-1}$). For example, silica glass has only Band 1, but $Na_2O$–CaO–$SiO_2$ glasses have all three bands [6]. Determination of total water content requires a summation of all three bands. However, Band 3 is not easily detected in an original spectrum, because various absorption bands due to network formers (i.e., Si–O–Si, P–O–P, B–O–B...) are overlapped. For the metaphosphate glasses of this study, the $OH^-$ can be assumed to be bound to chain ends [7]; this means the metaphosphate glasses have only one band due to water, although the peak wavenumber of the band depends on the glass composition. Therefore, the water content was taken to be proportional to the absorbance of "Band 2" in the normalized spectra.

Generally, the electrical conductivity ($\sigma$) depends on temperature as shown by eq. (3) and on the number ($N$) of electric charge carriers per unit volume, eq. (4);

$$\sigma = \sigma_0 \exp(-E_{dc}/RT), \tag{3}$$

$$\sigma = Ne\mu, \tag{4}$$

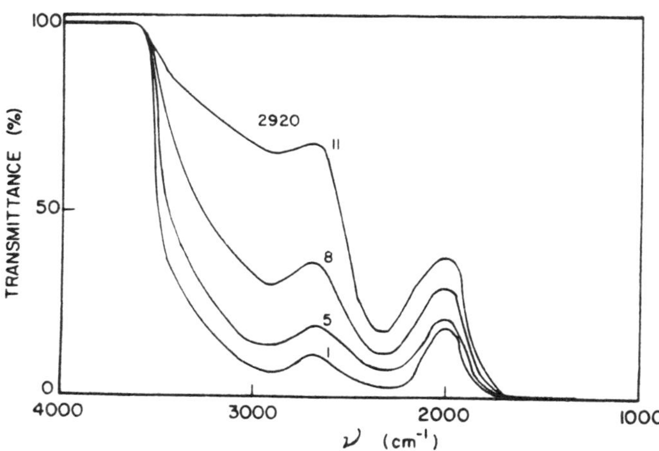


Fig. 1. Examples of IR spectra of a calcium metaphosphate glass plate specimen (The number indicated in the figure denotes the sample number in table 1) (0.5 mm thick specimen).

Table 1
Electrical conductivity (dc) data of alkaline earth metaphosphate glasses

| Alkaline earth metal | OH band (peak) ($cm^{-1}$) | $[H^+]$ content (mol/l) | log $\sigma$ at 417 K (mho $cm^{-1}$) | log $\sigma_0$ | log $\Lambda_0$ at 417 K | $E_{dc}$ (kcal/mol) |
|---|---|---|---|---|---|---|
| Be- | 3260 | 0.28 | −16.01 | −2.55 | −14.90 | 25.66 |
| Mg- | 3190 | 0.16 | −14.55 | 1.17 | −12.96 | 29.96 |
| Ca-1 | 2920 | 0.46 | −11.67 | 2.06 | −10.99 | 26.07 |
| Ca-2 | 2920 | 0.42 | −11.76 | 2.18 | −11.00 | 26.53 |
| Ca-3 | 2920 | 0.37 | −11.88 | 2.16 | −11.02 | 26.64 |
| Ca-4 | 2920 | 0.36 | −11.90 | 2.26 | −11.00 | 26.87 |
| Ca-5 | 2920 | 0.34 | −11.90 | 2.13 | −10.96 | 26.73 |
| Ca-6 | 2920 | 0.29 | −12.02 | 2.37 | −10.94 | 27.10 |
| Ca-7 | 2920 | 0.26 | −12.09 | 2.18 | −10.91 | 27.20 |
| Ca-8 | 2920 | 0.21 | −12.38 | 2.13 | −11.02 | 27.66 |
| Ca-9 | 2920 | 0.16 | −12.69 | 2.18 | −11.08 | 28.24 |
| Ca-10 | 2920 | 0.15 | −12.78 | 2.16 | −11.11 | 28.46 |
| Ca-11 | 2920 | 0.07 | −13.32 | 1.75 | −11.06 | 29.04 |
| Sr- | 2890 | 0.10 | −12.86 | 2.57 | −10.86 | 29.41 |
| Ba- | 2800 | 0.24 | −11.06 | 2.19 | −9.82 | 25.25 |

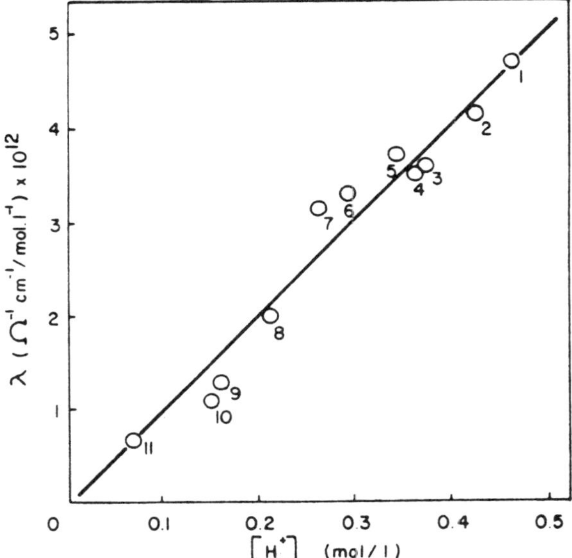


Fig. 2. Plot of equivalent conductivity versus proton concentration of calcium metaphosphate glasses (the number indicated in the figure denotes the sample number in table 1) (at 417 K).

where $E_{dc}$ is the apparent activation energy for dc electrical conduction; $R$ is the gas constant; $T$ is the temperature in degrees Kelvin; $\sigma_0$ is the pre-exponential term; $e$ is the electrical charge; and $\mu$ is mobility. Table 1 gives the electrical conductivity data of alkaline earth metaphosphate glasses in relation to the water content of the glasses. Fig. 2 plots the equivalent conductivity, $\lambda$ (mho $cm^{-1}$/mol $l^{-1}$) at 417 K versus proton concentration $[H^+]$ of calcium metaphosphate glasses, where $\lambda$ is $\sigma/[H^+]$, i.e., conductivity per one mole proton per liter of glass. Eq. (5) was experimentally obtained from fig. 2:

$$\lambda \times 10^{12} = 10.43[H^+] - 0.10. \tag{5}$$

Therefore, the conductivity is expressed by eq. (6),

$$\sigma_{417} = A_0[H^+]^2, \tag{6}$$

where $\sigma_{417}$ is conductivity at 417 K and $A_0$ is a constant of $1.04 \times 10^{-11}$ (av.) (mho $cm^{-1}$ $mol^{-2}$ $l^2$) for calcium metaphosphate glasses.

By combining eq. (6) with eq. (4), we see that for calcium metaphosphate glass the mobility of protons in the glass increases linearly with increasing proton concentration. The dependence of $E_{dc}$ on $[H^+]$ is obtained by combining eq. (3) with eq. (6):

$$E_{dc} = 2.303\,RT\left(\log A_0 - \log[H^+]^2\right). \tag{7}$$

The values of $\sigma_0$ and $A_0$ for calcium metaphosphate glasses are essentially constant and independent of the water content or the conductivity, as shown in table 1. Substituting into eq. (7) the average experimental values (table 1) of

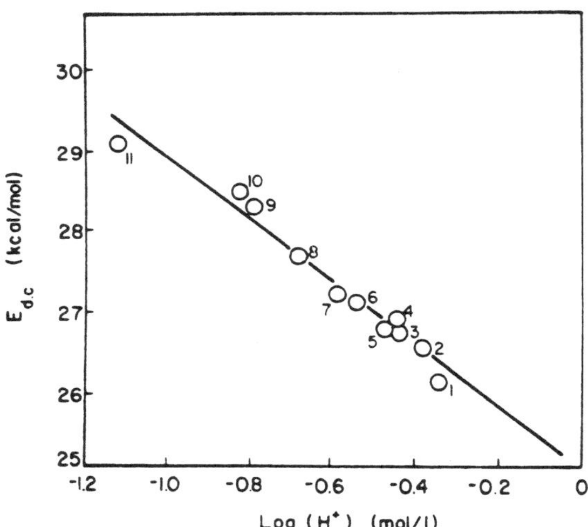


Fig. 3. Relation between apparent activation energy for electrical conduction ($E_{dc}$ and proton concentration ($[H^+]$) of calcium metaphosphate glasses. ($\nu = 2920$ $cm^{-1}$) (the number indicated in the figure denotes sample number in table 1).

$\log \sigma_0 = 2.14$, $\log A_0 = 11$, and $T = 417$ K for calcium metaphosphate glasses, which have the same OH absorption frequency (2920 $cm^{-1}$) and different $[H^+]$, eq. (8) is obtained:

$$E_{dc} = 25.0 - 3.8 \log[H^+]. \tag{8}$$

Fig. 3 shows the $E_{dc}$ versus $\log[H^+]$ plot.

Thus, $E_{dc}$ decreases linearly with $\log[H^+]$ when the O–H bonding energy is the same (i.e., when $\nu$ is the same).

With the exception of the calcium metaphosphate series, the other metaphosphate glasses were not produced with various amounts of water. However, the value of $A_0$ in table 1 was calculated by assuming that eq. (6) holds for all the metaphosphate glasses. As shown in table 1, it is apparent that $A_0$ and $\sigma_0$ increase with decreasing peak wavenumber of the OH absorption band. That is to say, the weaker the O–H bonding is (or the stronger the OH...O hydrogen bonding is), the more mobile are the protons in the glasses. The degree of proton mobility in the alkaline earth metaphosphate glasses is of the order of Ba > Sr > Ca > Mg > Be. The relationship between $E_{dc}$ and $\nu$ must be discussed for various glasses which have the same $[H^+]$, because $E_{dc}$ depends on $[H^+]$ as can be seen for calcium metaphosphate glasses [eq. (8)].

It was confirmed that the so-called Ohm's law holds in this experiment (applied voltages: 9 to 54 V. See fig. 4). Some investigators may have doubts that the electric carrier in these metaphosphate glasses is not the proton but

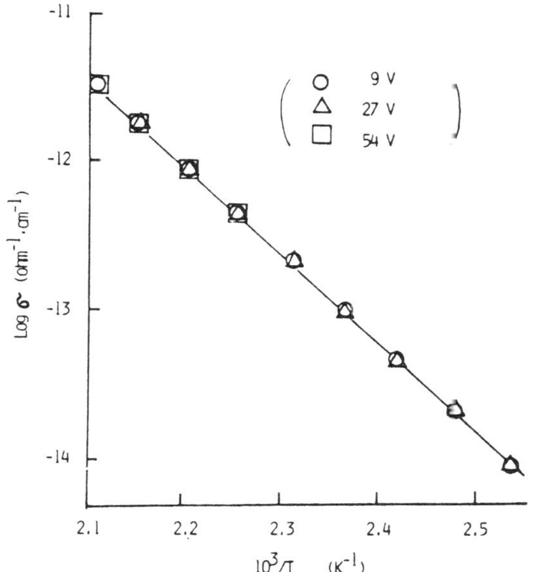


Fig. 4. Plot of conductivity versus $1/T$ for a calcium metaphosphate glass when different voltages are applied.

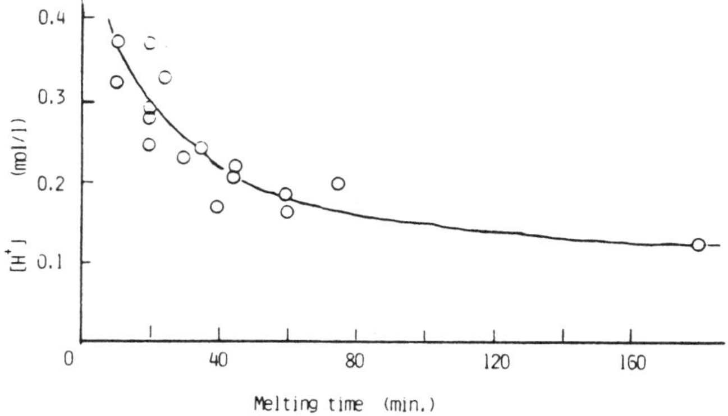


Fig. 5. Plot of melting time versus water content of $CaO \cdot P_2O_5$ glasses (melting temperature 1300°C).

alkali impurities, on the basis of the fact that ionic conduction by protons in solids is not a well-established phenomenon of the same general nature as conduction by alkali ions as seen in a review by Bruinink [9]. However, suspicion is not a fact. In these phosphate glasses, neither alkali ion impurities nor alkaline earth ions are electric carriers.

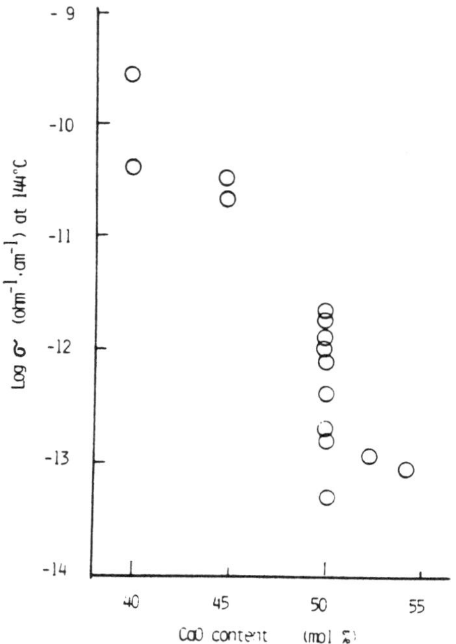


Fig. 6. Plot of conductivity versus CaO content in $CaO-P_2O_5$ glasses.

Fig. 5 is an example of calcium metaphosphate glasses, which shows how proton concentration decreases with increasing melting time; the content of alkali impurities, if present, should be essentially constant through the series of the specimen, because the glasses were prepared by using the same raw materials. The content of alkali impurities of the calcium metaphosphate glasses was determined by flame photometry as $Li^+$ (= 0.007 mol/l) < $Na^+$ (= 0.010 mol/l) ≫ $K^+$ (0.001 mol/l), all of these alkali contents are much less than that of $H^+$. Fig. 6 shows that $Ca^{2+}$ is not a carrier in the $CaO-P_2O_5$ glasses; the different values of the conductivity at the same CaO content of 50 mol% are caused by the different amounts of $H^+$ in the glasses, as can be seen in fig. 2. Doremus [10] estimated that the mobility of $Na^+$ is $10^4$ times that of $H^+$ in silica glass. This may be true in silica glass, because O–H bonding in silica glass is very strong; in ref. 11 we showed that log $A_0$ decreases linearly with increasing $\nu$ and that the mobility of a proton in a certain phosphate glass is approximately estimated to be $10^{10}$ times that in silica glass. Protons in phosphate glasses are very mobile because of the strong hydrogen bonding and weak O–H bonding.

## 4. Conclusions

These results show that it is possible to control the proton conductivity in alkaline earth metaphosphate glasses by varying the extent of dehydration during melting. The conductivity of calcium metaphosphate glasses is directly proportional to the square of the proton concentration as shown in eq. (6). The apparent activation energy for the conduction for calcium metaphosphate glasses decreases with increasing proton concentration as shown in eq. (8). Mobility of protons in alkaline earth metaphosphate glasses increases with decreasing O–H bonding strength or with increasing OH...O hydrogen bonding in the glasses. The value of $\sigma_0$ and $A_0$ seems to depend on $\nu$ but not on proton concentration.

The authors thank the Center of Excellence in Engineering Program at the University of Florida for partial financial support (YA) while preparing this paper. One of the authors (LLH) also acknowledges support of AFOSR Contract # F49620-80-C-0047.

## References

[1] L.L. Hench and H.F. Schaake, in Introduction to Glass Science, eds., D. Pye, H. Simpson, W. LaCourse (Plenum, New York, 1972) p. 583.
[2] K. Hughes, J.O. Isard and G.C. Milnes, Phys. Chem. Glasses 9 (1968) 43.
[3] M. Schwartz, Thesis, Rensselaer Polytech. Inst., Troy, NY (1969).
[4] E. Gough, J.O. Isard and J.A. Topping, Phys. Chem. Glasses 10 (1969) 89.
[5] H. Namikawa and Y. Asahara, Yogyo Kyokai Shi (J. Ceram. Soc. Japan) 74 (1965) 205.

[6] H. Scholze, Glastech. Ber. 32 (1959) 81.
[7] A. Naruse, Y. Abe, and H. Inoue, Yogyo Kyokai Shi (J. Ceram Soc. Japan) 76 (1968) 36.
[8] Y. Abe and D.E. Clark, to be submitted to J. Mater Sci.
[9] J. Bruinink, J. Appl. Electrochem. 2 (1972) 239.
[10] R.H. Doremus, J. Electrochem. Soc. 115 (1968) 181.
[11] Y. Abe, Y. Ohta, H. Hosono and L.L. Hench, to be submitted to Sol. St. Commun.

# PHOTO- AND THERMO-COLORING OF REDUCED PHOSPHATE GLASSES

Y. Abe and R. Ebisawa

*Nagoya Institute of Technology, Inorganic Materials*
*Showa-ku, Nagoya 466, Japan*

D.E. Clark and L.L. Hench

*University of Florida*
*Gainesville, FL 32611*

Various reduced phosphate glasses show striking phenomenon. In the as-cast glasses which were made with a higher cooling rate from the melts, phosphorus may be atomically dispersed or it may form very fine particles; this leads to transparent and colorless glasses when the glasses contain no coloring agents such as transition metal ions. However, phosphorus associates into colloidal red phosphorus (∿50 nm) to give a reddish color to the glasses, when the glasses are subjected to a heat-treatment at moderately high temperatures (e.g., 400°-600°C) (1-3). This phenomenon is known as "striking." The colored glasses are bleached to be almost transparent and colorless again, when they are subjected to a heat-treatment above 600°C and subsequent quenching. These bleached glasses are referred to as PTC-RP glasses (3), since they were found to be Photo-and Thermo-Colorizable Reduced Phosphate glasses. They are easily red-colored either by sunlight (2) or UV light at room temperature or by heating above 200°C (3). Thermo-coloring of reduced glasses having composition of $3K_2O\cdot 12B_2O_3\cdot 69P_2O_5$ and $6K_2O\cdot 22Al_2O_3\cdot 72P_2O_5$, both of which have glass transition temperature >600°C have been reported (3).

In the present paper, only a simple composition, as an example of PTC-RP glass, is discussed. The dependence of sensitivity on the irradiation light wavelength is described in this paper.

## Sample Preparation

The following three kinds of glasses were prepared.

(a) Glass reduced by silicon (CP-RS): A mixture of di-hydrogen calcium phosphate, $Ca(H_2PO_4)_2\cdot H_2O$ and metallic silicon powder were melted in an alumina crucible at 1200°-1250°C for 1 hour and poured on a cold graphite plate. The silicon was 0.15 mole per mole of $P_2O_5$. (Formula: $CaO\cdot P_2O_5\cdot 0.15SiO_2$)

(b) Glass reduced by ammonium salt (CP-RN): A mixture of

$Ca(H_2PO_4)_2$ $H_2O$, $CaCO_3$, and $NH_4H_2PO_4$ were used as raw materials. 20% of $P_2O_5$ of the glass was supplied by using $NH_4H_2PO_4$. Glasses were prepared in a similar way to CP-RS. (Formula: CaO $P_2O_5$)
(c) Glass not reduced (CP-OS): Glasses were prepared in the same way as CP-RS except that $SiO_2$ was substituted for Si in the batch. (Formula: CaO $P_2O_5$ 0.15$SiO_2$)
All the glasses were annealed and all reagents were of reagent grade.

## Striking (Coloring by Colloid Formation)

Reduced phosphate glasses of which as-cast glasses are transparent and colorless become tinged with red to yellow depending on the reheating time and temperature. This "striking phenomenon" is due to the formation of colloidal red phosphorus (1).

The specimens of CP-RS and CP-RN, both of which were prepared under a reducing condition, exhibit striking phenomenon but CP-OS does not. An example of the optical transmission curves of CP-RN glass specimen heated at 580°C is given in Figure 1, which shows similar changes to that for CP-RS previously reported (2), namely, that the absorption edge at shorter wavelength sides shifts to longer wavelength with increasing time of reheating. Thus, the specimens become tinged with red. The coloring is saturated for 30-50 hours at 580°C.

## Photo-Coloring (Photochromism)

The bleached specimens (PTC-RP glasses) were prepared by heating the saturated-red colored specimens at ∿600°C for several minutes and subsequent quenching. Figure 2 gives an example of the transmission curves in coloring of the bleached CP-RN glass specimen exposed to solar rays. The bleached specimen of CP-RS (2) exhibited trends similar to that of CP-RN. Figure 3 shows the dependence of coloring sensitivity bleached CP-RS glass on the wavelength of the irradiating light, where $\Delta e_{40}$ in the ordinate gives a parameter of coloring sensitivity. This parameter is the difference between the photon energy (ev) corresponding to wavelength at 40% of the transmission curve of a bleached glass before and after the irradiation. As is shown, it was found to be most sensitively colored by irradiating with light of ∿350 nm but it is not colored by a wavelength longer than ∿500 nm.

As an example of change in coloring of a PTC RP specimen of CP-RS glass by the irradiation with a given wavelength (253.7 nm), a plot of $\lambda_{\frac{1}{2}\cdot T}$ is used as a measure of the absorption edge, indicating a wavelength corresponding to half of the maximum transmission of the spectrum (approximately a wavelength at a transmission of 40%).

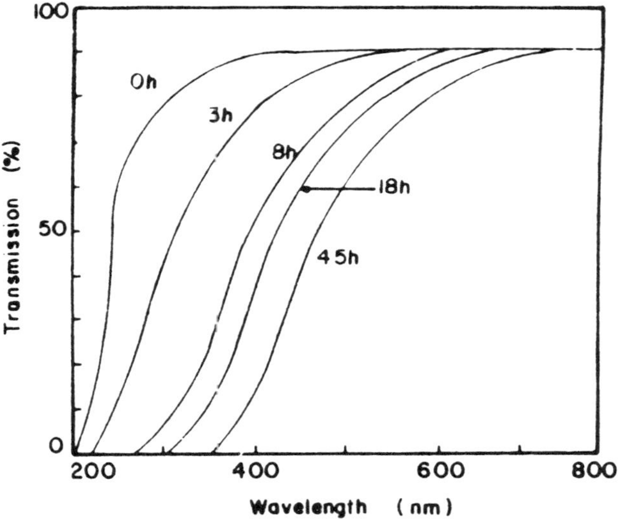


*Figure 1. Transmission curves of CP-RN glass heated at 580°C (striking; 1 mm thick specimen).*

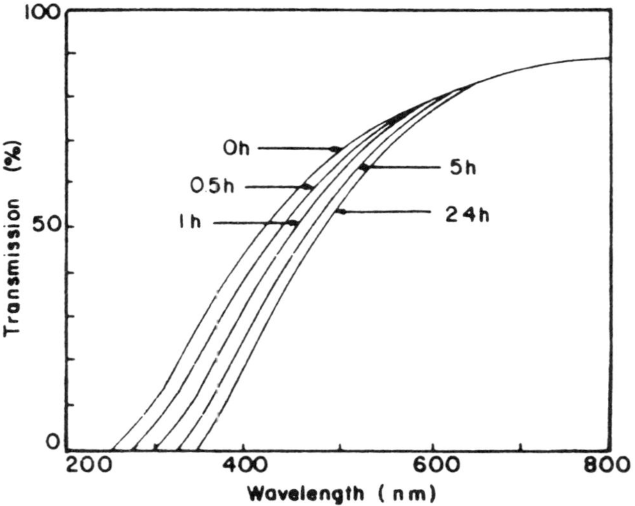


*Figure 2. Transmission curves of a bleached CP-RN glass exposed to solar rays (photo-coloring; 1 mm thick specimen).*

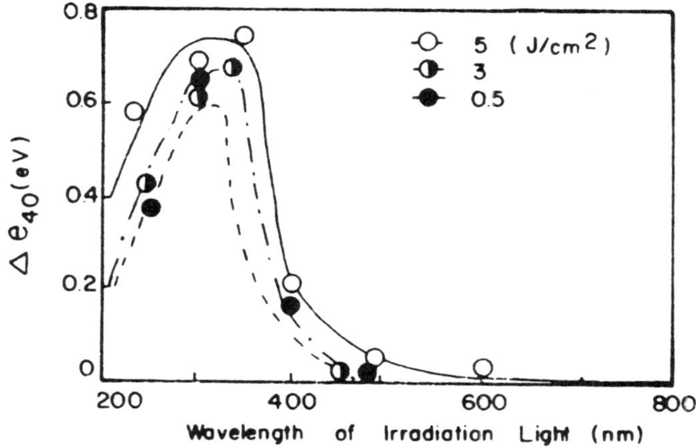


*Figure 3. Dependence of coloring sensitivity of a bleached CP-RS glass on the wavelength of irradiation light (1 mm thick specimen).*

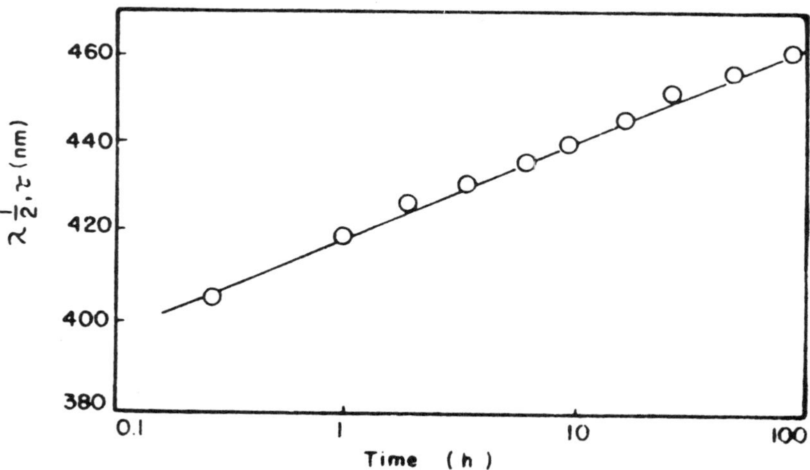


*Figure 4. Coloring of a bleached CP-RS specimen by irradiation light (253.7 nm 750 μW cm$^{-2}$, 1 mm thick specimen).*

Thermo-Coloring (Thermochromism)

As previously reported on some glasses of different composition (3), PTC-RP glass is colored also by heating above 200°C. For convenience, we define a coloring paramter $\alpha$ in eq. [1].

$$\alpha = (\lambda_{1/2 \cdot \tau} - \lambda_{1/2 \cdot i})/(\lambda_{1/2 \cdot s} - \lambda_{1/2 \cdot i}) \ldots\ldots\ldots\ldots\ldots [1]$$

where, $\lambda_{1/2 \cdot \tau}$ is $\lambda_{1/2}$ after heating for time $\tau$,

$\lambda_{1/2 \cdot i}$ is $\lambda_{1/2}$ before heating (initial specimen),

$\lambda_{1/2 \cdot s}$ is $\lambda_{1/2}$ of a well colored (saturated) specimen.

An apparent activation energy for the thermo-coloring was estimated to be ∿35 Kcal/mol, according to the method described previously (3). This value is almost the same as that of the other glasses reported by Abe et al. (3).

Determination of Elemental Phosphorus Formed in the Reduced Glasses

The permanganometric method developed by Venugopalan (4) was modified in our laboratories. It was found that the reduced glass contained ∿0.2 wt % of elemental phosphorus.

Mechanism of Coloring and Bleaching

The coloring and bleaching phenomena are caused by changes in molecular configuration of colloidal phosphorus formed in the glasses. The $P_4$ molecule (liquid or white phosphorus) is colorless. It polymerizes thermally or photochemically to the ···–P(P|P)P–··· configuration which gives a reddish color; it dissociates into $P_4$ molecule again when melted.

Literature Cited:

1. A. Naruse, Y. Abe, J. Ceram. Soc. Japan, 1965, 73, 253-58.
2. Y. Abe, R. Ebisawa, A. Naruse, J. Am. Ceram. Soc., 1976, 59 453-54.
3. Y. Abe, K. Kawashima, S. Suzuki, J. Am. Ceram. Soc., 64, 206-09 (1981).
4. M. Venugopalan, K. U. Matha, Z. Anal. Chem., 1956, 151, 262-68.
M. Venugopalan, K. J. George, Bull. Chem. Soc. Japan, 1957, 30, 51-53.

# DETERMINATION OF COMBINED WATER IN GLASSES

**Yoshihiro Abe**

*Department of Inorganic Materials, Nagoya Institute of Technology*
*Gokiso-cho, Showa-ku, Nagoya 466, Japan*

**David E. Clark**

*Department of Materials Science and Engineering, Ceramics Division*
*College of Engineering*
*University of Florida, Gainesville, FL 32611*

The presence of water in glass significantly affects certain of its properties such as electrical conductivity, index of refraction, density and viscosity. To evaluate quantitatively the effects of water on the properties of glass, both the amount and structural positions of the water must be known. The precise measurement of water content is very difficult and time consuming. A simple method proposed by Griffith[1] for alkali-phosphate glasses involves determining the weight loss on ignition of a mixture of fine glass powders and ZnO. The change in weight upon ignition is assumed to be equivalent to the water content of the glass. Although this method can be used for glasses containing large amounts of water, it is not applicable to many commercially important glasses which contain less than 0.1wt%. Furthermore, this method does not provide any direct information on the structural position of the water in the glass.

The molar absorption coefficient ($\varepsilon_{H_2O}$ or $\varepsilon_{OH}$) is often used for the determination of water in glass. $\varepsilon_{H_2O}$ is obtained by plotting the absorbance due to water versus the amount of water in the glass. Absorbance is commonly measured by infrared spectroscopy using a flat glass plate specimen. The water content can be measured directly by weighing the glass after heating inside a vacuum chamber. The relationship between the absorbance (E) and the water content (c) is linear and the slope is equal to $\varepsilon$ for a constant sample thickness (d). Once E has been obtained for a specific glass composition it can be used for determining the amount of water in other glass sheets with the same composition. The amount of water is given by:

$$c = \frac{E}{\varepsilon d} \quad (1)$$

c - water content ($H_2O$, mol/ℓ).

E - absorbance at absorption peak due to the OH vibration determined from the ir spectrum.

ε - molar absorption coefficient (as $H_2O$, ℓ/mol-cm).

d - thickness of specimen (cm)

The values of ε, which depend on the wavenumber of absorption band due to water, have been determined by many investigators. Unfortunately, these do not always agree with each other. In this paper, we use the value of $\varepsilon_{H_2O}$ when given and if $\varepsilon_{OH}$ was used in original papers, this was converted to $\varepsilon_{H_2O}$. These data are plotted in Fig. 1. The main absorption bands due to water[2] are usually detected in the range 2.75 to 2.95 μm (Band 1), 3.35 to 3.85 μm (Band 2), and 4.25 μm (Band 3). Band 1 is due to a free OH, Band 2 is due to weak hydrogen bonding between OH and a nonbridging oxygen, and Band 3 is due to the strongest hydrogen bonding of OH to a nonbridging oxygen. Scholze[2] did a systematic study of $\varepsilon_{H_2O}$ for various silicate glasses and pointed out that the total water content should be calculated by the sum of Band 1, Band 2, and Band 3. Silica glass has only one band (Band 1) but silicate glasses containing network modifying ions usually have all three bands which are often broad and superimposed. Scholze[2] also showed that the distribution of the absorbing species among the three bands remains constant as the concentration of water in the glass changes when the basic glass composition remains the same. Therefore, the determination of the total water concentration by ir absorption can be based on a single band. He showed that for a $16 Na_2O \cdot 10CaO \cdot 74 SiO_2$ glass the practical molar absorption coefficient ($\varepsilon_{pract}$) is 41 ℓ/mol·cm (6 in Fig. 1) at 2.85 μm (3500 $cm^{-1}$) although his $\varepsilon_{H_2O}$ at the same wavenumber is 75 ℓ/mol·cm. Thus, it is easy to understand why different $\varepsilon_{H_2O}$ values at the same wavenumber have been reported by different investigators. In Fig. 1, the data points labelled 1[5] and 2[3] are $\varepsilon_{pract}$ as pointed out by Götz.[3] Data points 4[7] and 5[3] should also be regarded as $\varepsilon_{pract}$ since the glass compositions are complicated and bands due to water are somewhat superimposed. The value $\varepsilon_{H_2O}$ and $\varepsilon_{pract}$

at 3660 $cm^{-1}$ for silica glass are identical because it has only the one band. Day and Stevels[4] obtained a straight line by plotting the decrease in absorbance at 2920cm$^{-1}$ versus the weight loss of $Na_2O \cdot P_2O_5$ glass after it was heated. Data point 7[4] was calculated by assuming that the weight loss is caused only by the loss of water from this glass. Data point 8 is Franz's[9] $\varepsilon_{OH}$ value changed to $\varepsilon_{H_2O}$ by Pasteur.[8] Figure 1 suggests that $\varepsilon$ depends fundamentally on $\nu$ but not on glass composition.

The total water content for the $Na_2O$-$CaO$-$SiO_2$ glass can be determined by using only one coefficient ($\varepsilon_{pract}$) as a representative $\varepsilon$, instead of using the respective $\varepsilon_{H_2O}$ at Band 1, Band 2, and Band 3. However, in order to determine the respective contributions of water content due to Band 1, Band 2 and Band 3 separately, the infrared absorption bands must be resolved. We propose that for the approximate determination of total water content $\varepsilon_{pract}$ (dotted line, Fig. 1) can be used, and that for the separate determination of water due to Band 1, Band 2 and Band 3, individual $\varepsilon_{H_2O}$ (solid line, Fig. 1) should be used. Investigators determining the amount of water in glasses should specify which $\varepsilon$ value they are reporting in order to prevent confusion.

$$C_{precise} = \frac{1}{d}\left(\frac{E\ (Band\ 1)}{\varepsilon\ (Band\ 1)} + \frac{E\ (Band\ 2)}{\varepsilon\ (Band\ 2)} + \frac{E\ (Band\ 3)}{\varepsilon\ (Band\ 3)}\right) \quad (2)$$

$$C_{Approx.} = \frac{1}{d} \quad \frac{E}{\varepsilon_{pract}} \quad (3)$$

## Acknowledgements

The authors thank the Center of Excellence in Engineering program at the University of Florida for partial financial support while preparing this paper.

References


1. E. J. Griffith and C. F. Callis, "Structure and Properties of Condensed Phosphates, XV. Viscosity of Ultraphosphate Melts," J. Am. Chem. Soc., 81, 833 (1959).

2. H. Scholze, "Structure of Water in Glasses. 1. Effect of Dissolved Water in Glass on the Infrared Spectra and Quantitative Infrared Spectroscopic Determination of Water in Glasses," Glastech. Ber. 32, 81-88 (1959).

3. J. Götz and E. Vosáhlová, "Contribution to Quantitative Determination of Water Content of Glass from the Infrared OH Bands," Glastech. Ber. 41 47-55 (1968).

4. D. E. Day and J. M. Stevels, "Internal Friction of $NaPO_3$ Glasses Containing Water," J. Non-Cryst. Solids 11 459-470 (1973).

5. A. D. Pearson, G. A. Pastear and W. R. Northover, "Determination of the Absorptivity of OH in a Sodium Borosilicate Glass," J. Mater. Sci., 14 869-872 (1979).

6.a. D. M. Dodd and D. B. Fraser, "Optical Determinations of OH in Fused Silica," J. Appl. Phys. 37, 3911 (1966).

b. A. J. Moulson and J. P. Roberts, "Water in Silica Glass," Trans. Brit. Ceram. Soc. 59, 388-399 (1960).

7. R. F. Bartholomew, B. L. Putler, H. L. Hoover, and C. K. Wu, "Infrared Spectra of a Water-Containing Glass," J. Am. Ceram. Soc., 63 481-485 (1980).

8. G. A. Pasteur, "Optical Determination of OH in $B_2O_3$ Glass," J. Am. Ceram. Soc., 56 548 (1973).

9. H. Franz, "Investigation on the Acid-Base Conditions in Oxide Melts: I," Glastech. Ber. 38 54-59 (1965).

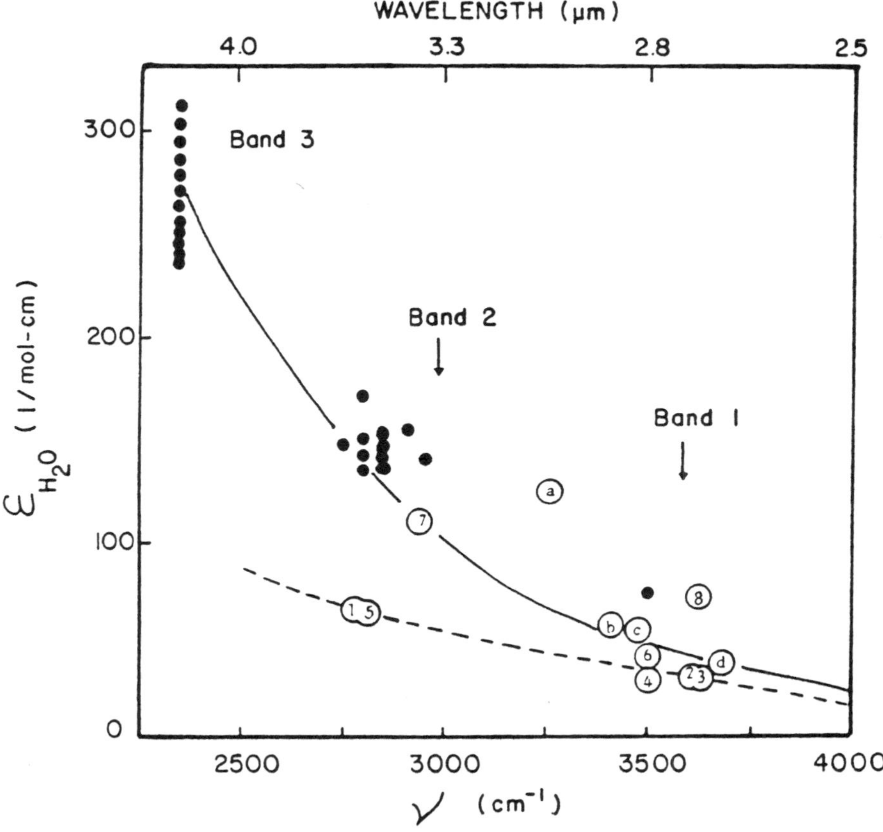


Figure 1. $\epsilon_{H_2O}$ vs. $\nu$ for several materials.

Notation for Figure 1

● - silicate glasses (Ref. 2)
a - ice
b - water at 3°C
c - water at 60°C
d - water in $CCl_4$
1 - 20 $Na_2O \cdot 35B_2O_3 \cdot 45SiO_2$ glass (Ref. 5)
2 - 15$Na_2O \cdot 9.5CaO \cdot 72.25SiO_2$ glass in wt % (Ref. 3)
3 - $SiO_2$ glass (Ref. 6)
4 - 10.8$Na_2O \cdot 3.0K_2O \cdot 7.8ZnO \cdot 1.3Al_2O_3$ 76.9$SiO_2$ glass (Ref. 7)
5 - 15$Na_2O \cdot 9.5CaO \cdot 72.25SiO_2$ glass (Ref. 3)
6 - $\epsilon_{pract}$ for 10$Na_2O \cdot 16CaO \cdot 74SiO_2$ glass in wt % (Ref. 2)
7 - $Na_2O \cdot P_2O_5$ glass (Ref. 4)
8 - $B_2O_3$ glass (Ref. 8)

# NOISE IN N-TYPE α-SILICON CARBIDE

J.S. Kim, G. Bosman, and C.M. Van Vliet

*Department of Electrical Engineering, College of Engineering*
*University of Florida*
*Gainesville, FL 32611*

## I. Introduction

Silicon carbide [1] exhibits the phenomenon of polytypism [2], in which the same chemical compound containing Si and C atoms crystallizes into different crystallographic modifications known as polytypes. These polytypes are all similar in the plane perpendicular to the symmetry axis but differ from each other in the stacking sequence and cycle in the direction of the symmetry axis.

It is well-known [3-9] that various polytypes of SiC appreciably differ in semiconducting properties such as energy bandgap, electron mobility, luminescence and absorption spectra, etc. With a view to establishing a simple relationship [5,7-9] between polytype structures and semiconducting properties, the concept of a one-dimensional superlattice [10] seems appropriate to explain a wide range of experimental facts regarding these differences.

However, virtually all crystals of SiC are known to exist in a syntactic coalescence of different polytypes. The different polytypic modifications usually appear to form together under nearly the same conditions. This phenomenon was initially noted by Baumhauer [11] who described crystals which changed in structures along the c-axis. Later, extensive studies were directed to understand the formation of lamellae in connection with possible growth kinetics [12]. However, due to the ambiguity in defining the large number of factors which compete in the growth process, the controlled growth of single polytypic structures is still not possible at present. Therefore one-dimensional inhomogenities have often given rise to conflicting evidence in defining the anisotropic nature of a given polytype [5,13], which can be predicted by its superlattice structure. Consequently, although there has been considerable interest in meaningful applications of SiC to electronic devices [14], little is known about the detailed electrical conduction process in these composite SiC crystals.

In order to achieve a better understanding of the charge carrier transport properties which are closely associated with crystal structure, a noise analysis in n-type α-SiC crystals has been undertaken. The experimental results will be presented in the following section of this report. A review of the electronic band structure of polytypic SiC has also been prepared and will be included in next year's progress report.

## II. Experiment and Result

### 2.1 Preparation of sample

The sample used in this experiment was sliced from a thin platelet of α-SiC crystal, in which a lamellar structure was visually discernable, by using a diamond impregnated dicing wheel (Tempress 16852). The thickness of the blade was about 5.5 mils. For reducing the forces exerted on the crystal surface, we chose repetitive sweeps of the spindle to make a fine and gradual slice. The sample was prepared as a rectangular bar having well-developed crystal faces.

After etching away an oxide film in HF-HCl solution, tungsten ohmic contacts [15] were fabricated on both sides of the flat face of the crystal. The deposition of tungsten was made by a sputtering technique. Figure 1 shows the schematic of the sample structure mounted on an alumina plate with conductive patterns of gold. It consists of four terminals; two for carrying current and the other two for measuring the open circuit voltage developed across them.

### 2.2 Measurements of DC current-voltage characteristics

The DC current-voltage characteristics were measured at room temperature by employing different terminals. In order to avoid a rise in lattice temperature during the measurement a pulsed current was applied, especially at the high current regime. The results are shown in figures 2-4. Figure 2 shows the I-V characteristic along the symmetry axis. For low currents $I_{23}$ versus $V_{23}$ was linear with a resistance of 1200 ohms. At high currents it shows a $I_{23} \sim V_{23}^{1.5}$ characteristic. We also measured the $I_{14} + I_{23}$ versus $V_{23}$ characteristic under the condition of $V_{23} = V_{14}$. The characteristic remained unchanged. Figure 3 shows similar results from $I_{14}$ versus $V_{23}$ and $I_{41}$ versus $V_{32}$. In the low current region, the resistance was 1000 ohms for either case and it showed nonsymmetrical behavior with increasing current.

We observed a similar characteristic both in $I_{12}$ versus $V_{12}$ and $I_{34}$ and $V_{34}$ plots (Fig. 4). In this case within the range of current applied, no open circuit voltage was observed across the terminals at the opposite side. Also it was found that the superlinearity was less pronounced in the $I_{12}$ versus $V_{12}$ characteristic.

From these observations we can confirm that: (1) the superlinearity in I-V characteristics is ascribed in the lammelar structure: (2) the contact resistance is negligible: and (3) the crystal structure is not bilateral along the symmetry axis.

## 2.3 Impedance measurements

The impedance of the sample measured in the presence of several bias currents is shown in figures 5 and 6. At low frequencies a frequency independent characteristic was present, followed by an asymptotical approach to a $\omega^{-n}$ law for intermediate frequencies. Finally the SiC sample reached a current and frequency independent value for frequencies above 30 MHz. The total expression for frequency dependence of the impedance appears to be:

$$\left|Z(\omega)^2\right| = \left(\frac{dV}{dI}\right)^2 \sum_i \frac{1 + \omega^2\tau^2}{1 + \omega^2\tau_i} f(\tau_i) \qquad (1)$$

where $f(\tau_i)$ denotes a distribution of time constants, $\tau_i$, and dV/dI denotes the AC resistance for the zero frequency.

## 2.4 Noise measurements

The magnitude of the equivalent noise current for several bias current levels has been obtained in the frequency range from 1 Hz to 10 MHz. To cover this wide frequency range, several spectrum analyzers were employed. These are an HP-3582, GR 1925 multifilter accompanied by a GR 1926 rms detector, an HP 310A, and an HP 8557A. Each spectrum analyzer features a different method for the evaluation of the power spectrum in the available frequency ranges. The schematic of the noise measurement system, which includes the fundamental function of the spectrum analyzer, is shown in figure 7. In connection with this the equivalent noise circuit at the input of the amplifier is shown in figure 8. This measurement system enables us to calculate the absolute magnitude of the excess noise level due to the bias current.

The measurement procedure consists of three consecutive steps. The first step is the measurement of system noise which includes the thermal noise of equivalent resistance at the input of amplifier. This condition represents positioning both switches $S_1$ and $S_2$ at off-side as is shown in figure 7. Especially for high currents, it is necessary to take into account the change in the device impedance from the value at the equilibrium condition. To do this we replaced the whole input circuit by a resister equal to the real value of the parallel combination of $Z_{23}(I)$, $R_{cal}$ + 50 ohms, and $R_i$ as shown in figure 8.

The second step is the measurement of the excess noise in the presence of bias current through the sample. This condition represents positioning the switches $S_1$ at on-side and $S_2$ at off-side. At the last step the calibration signal from the white-noise generator is superimposed on the noise from the second step. In this case, the equivalent noise circuit is exactly the same as is shown in figure 8. Since each noise generator in figure 8 is uncorrelated with others, the three different noise levels measured from this procedure enable us to calculate the absolute magnitude of excess noise, $S_{i_{23}}$, in the presence of the bias current, $I_{14}$. Following the same idea, we also measured the noise directly across the DC-current carrying terminals. The results are shown in figures 9, 11, and 13.

The noise spectra of $S_{i_{12}}$ (figure 9) and of $S_{i_{14}}$ (figure 11) clearly show the 1/f type noise. The powers of the noise at several frequencies are plotted versus bias current in figures 10 and 12. It is noted that the expected behaviors for 1/f noise [16,17], $S_{i_{12}} \sim I_{12}^2$ and $S_{i_{14}} \sim I_{14}^2$, are well satisfied.

In contrast to the distinct 1/f type noise of $S_{i_{12}}$ and $S_{i_{14}}$, $S_{i_{23}}$ for $I_{14}$ shown in figure 13 appears to have several notable features. The 1/f slope for all current levels is seen for low frequencies. For low currents the spectrum is proportional to $1/f^{1/2}$ over the wide range of frequency, and finally it seems to turn into 1/f noise at the high frequencies. With increased bias current, the slope tends to be steeper and approachs 1/f type, and the appearance of a plateau can be seen clearly for the frequencies above 200KHz. The plot of $S_{i_{23}}$ versus $I_{14}$ at several frequencies (figure 14) does not represent the simple relationship of $S_{i_{23}} \approx I_{14}^2$. This simultaneous contribution of 1/f and g-r [18] type noises possibly gives rise to the distinct characteristics of $S_{i_{23}}$.

## III. Discussion

One of the major effects of the composite structure of polytypic SiC is simply the nonuniform distribution of mobile carriers along the c-axis direction. In a n-type crystal, the region with high electron concentration acting as a reservoir of electrons supplies all the electrons that the neighboring region with high resistivity demands for any applied filed. These injected electrons are drawn into that resistive media where they exercise possibly drift and diffusion, generation and recombination, and dielectric relaxation. Therefore, in addition to the majority carrier injection at the boundary, the electron conduction mechanism in that resistive region is supposed to control dominantly the current-voltage, the impedance, and the noise characteristics.

It is worth noting that usually SiC crystallizes into one of the three commonly occurring structures (6H, 15R, and 4H). The unit cell dimensions along the c-axis direction of these polytypes are relatively small. They are 15.12Å, 37.30Å, and 10.05Å respectively [1]. The bulk property of a polytype can be defined when the thickness of the polytypic layer is much larger than its own unit cell length. Therefore, there is a good reason to believe that 6H, 15R, and 4H satisfy this requirement. According to the principle of energy band line up in isotypic heterostructures [19,20], a proposed band structure of a SiC polytype-polytype junction at the equilibrium condition is shown in figure 15. The depletion region in the polytype with the higher bandgap may give rise to the large intensity of 1/f noise, which was observed in the spectra measured across the current carrying terminals. This type of noise is similar to the case of MOSFET structures [21].

However, considering the different results obtained from the four terminal measurements, the existence of energy barriers at polytype-polytype junctions may not cause substantial effects on the impedance to the charge carrier conduction along the c-axis direction.

According to the van Vliet's theory [22] on single injection diodes, there is a good evidence to associate the limiting factor in the conduction with the one-dimensional SCL flow in the presence of trapping. A quantitative comparison of the experimental data from impedance and noise measurements with the theory will be made in the near future. The results will be described in Jae Kim's Ph.D. thesis, which is presently in preparation.

References


1. The last survey in this field was published in 1974; R.C. Marshall, J.W. Faust, Jr., and C.E. Ryan, Silicon Carbide-1973, University of South Carolina Press, Columbia, S.C. (1974)
2. A.R. Verma and P. Krishna, Polymorphism and Polytypism in Crystals, Wiley, N.Y. (1966)
   P. Krishna and A.R. Verma, Phys. Stat. Sol. 17, 437 (1966)
3. W.J. Choyke, D.R. Hamilton, and L. Patrick, Phys. Rev. 133, A1163 (1964)
4. D.C. Barrett and R.B. Campbell, J. Appl. Phys., 38, 53 (1967)
5. G.A. Lomakina, Y.A. Vodakov, E.N. Mokhov, V.G. Oding, and G.F. Kholuyanov, Sov. Phys.-Solid St. 12, 2356 (1971)
6. T.S. Moss, G.J. Burrell, and B. Elliss, Semiconductor Opto-Electronics, Butterworths, London (1973)
7. G.B. Dubrovskii and A.A. Lepneva, Sov. Phys.-Solid St., 19,730 (1977)
8. G.B. Dubrovskii and E.I. Radovanova, Sov. Phys.-Solid St., 11,545 (1969)
9. G.B. Dubrovskii, A.A. Lepneva, and E.I. Radovanova, Phys. Stat. Sol. (b), 57, 423 (1973)
10. L. Esaki, and R. Tsu, IBM J. Res. Develop. 14, 61 (1970)

11. H. Baumhauer, Z. Kristallogr., 50 33 (1912) and Z. Kristallogr. 55, 249 (1915-1920)

12. W.F. Knippenberg, Phil. Res. Rept. 18, 161 (1963)

13. V.M. Lyubimskii, Sov. Phys.-Semicon. 13,609 (1979)

14. The detailed description of existing SiC junction devices can be found in P.B. Campbell and H.C. Chang, Silicon Carbide Junction Devices in Semiconductor and Semimetal vol. 7, part B, edited by R.K. Willardson and A.C. Beer, Academic Press, New York (1971). A SiC-BJT was fabricated with a moderate success and reported in W.V. Muench and P. Hoeck, Solid St. Electron., 21, 479 (1978)

15. R.N. Hall, J. Appl. Phys. 29, 914 (1958)

16. A. van der Ziel, Advances in Electronics and Electron Phys. 49, 225 (1979)

17. F.N. Hooge, T.G.M. Kleinpenning and L.K.J. Vandamme, Rep. Prog. PHys. 44, 479 (1981)

18. C.M. van Vliet and J.R. Fasset, Fluctuation Phenomena in Solids, edited by R.E. Burgess, Academic Press, New York (1965)

19. R.L. Anderson, Solid State Electron., 5, 341 (1962)

20. W.R. Frensley and H. Kroemer, Phys. Rev. 13, 16, 2642 (1977)

21. F.M. Klaasen, IEEE Trans., ED-18, 887 (1971)

22. C.M. van Vliet, A. Friedmann, R.J.J. Zijlstra, A. Gisolf, and A. van der Ziel, J. Appl. Phys., 46, 1814 (1975)

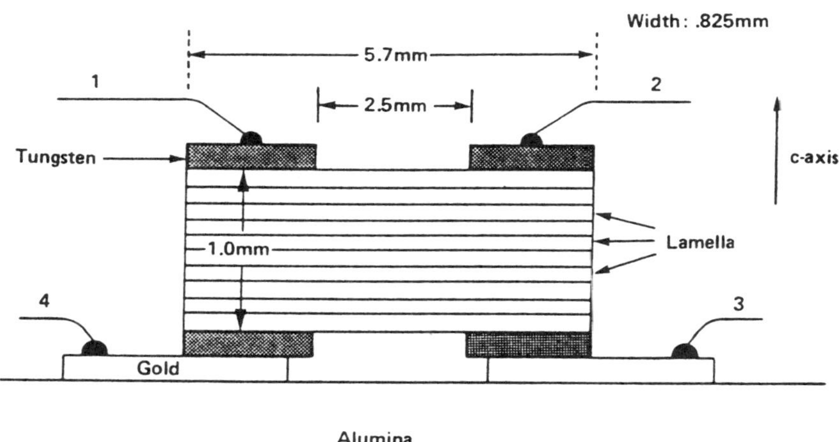


Fig. 1. The sample structure mounted on an alumina plate showing the lamellae along the c-axis direction and the arrangement of tungsten contacts.

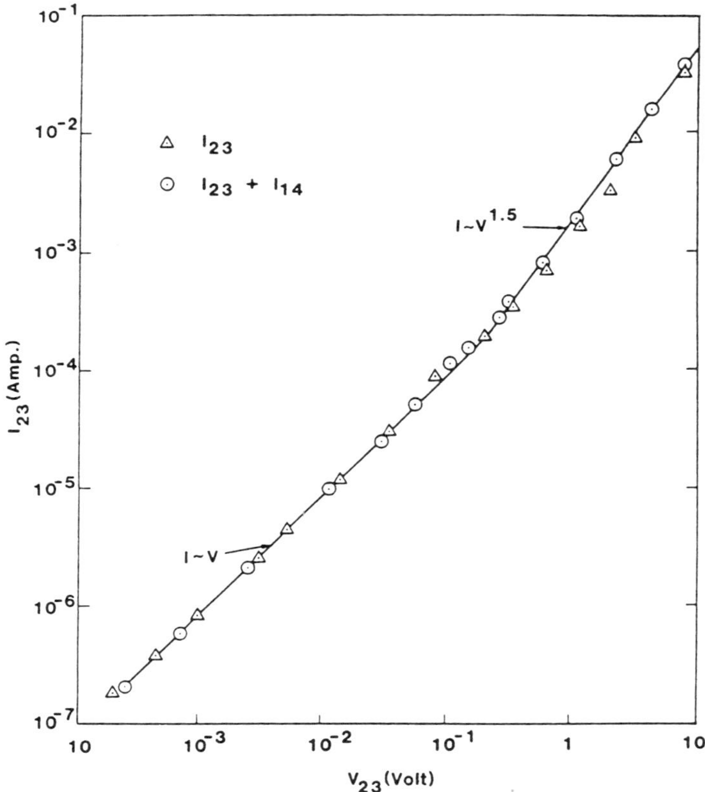


Fig. 2. $I_{23}$-$V_{23}$ and $I_{23} + I_{14} - V_{23}$ characteristics.

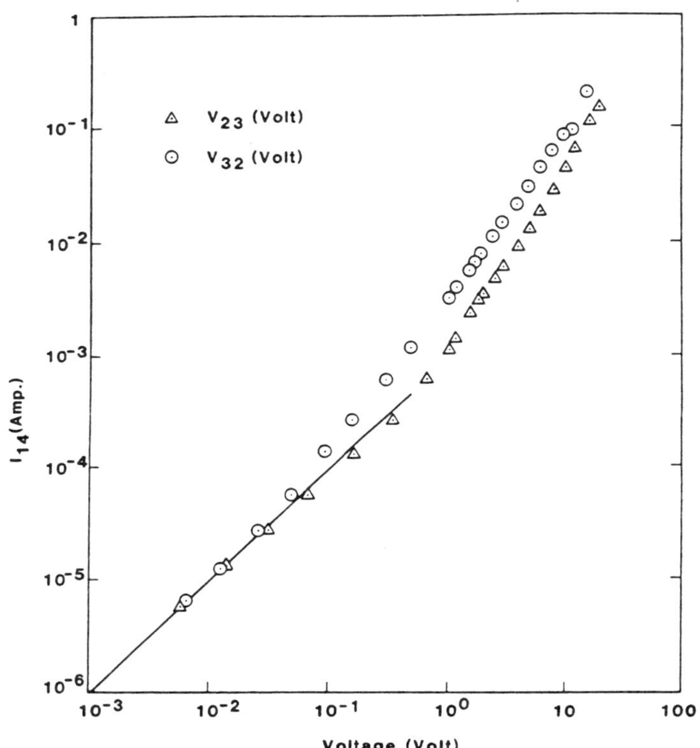


Fig. 3. $I_{14}$-$V_{23}$ and $I_{41}$-$V_{32}$ characteristics.

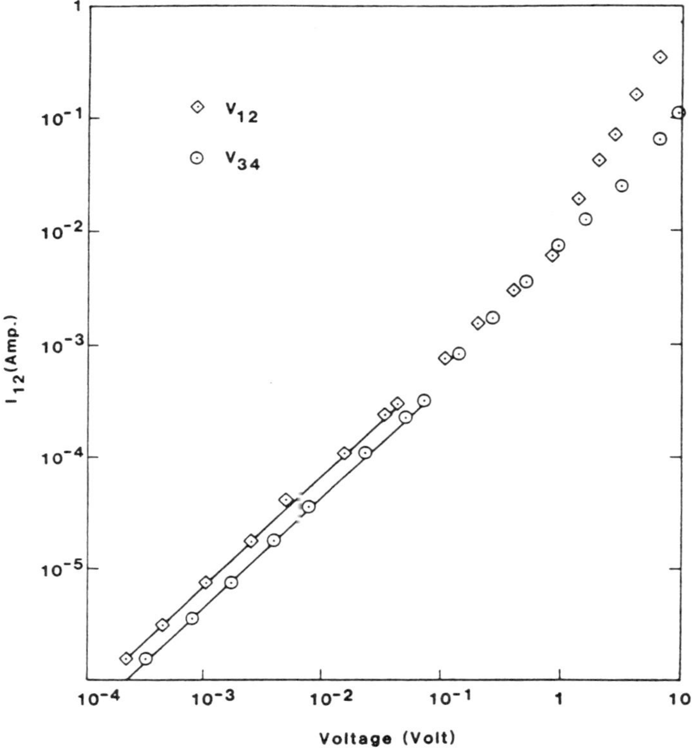


Fig. 4. $I_{12}$-$V_{12}$ and $I_{34}$-$V_{34}$ characteristics.

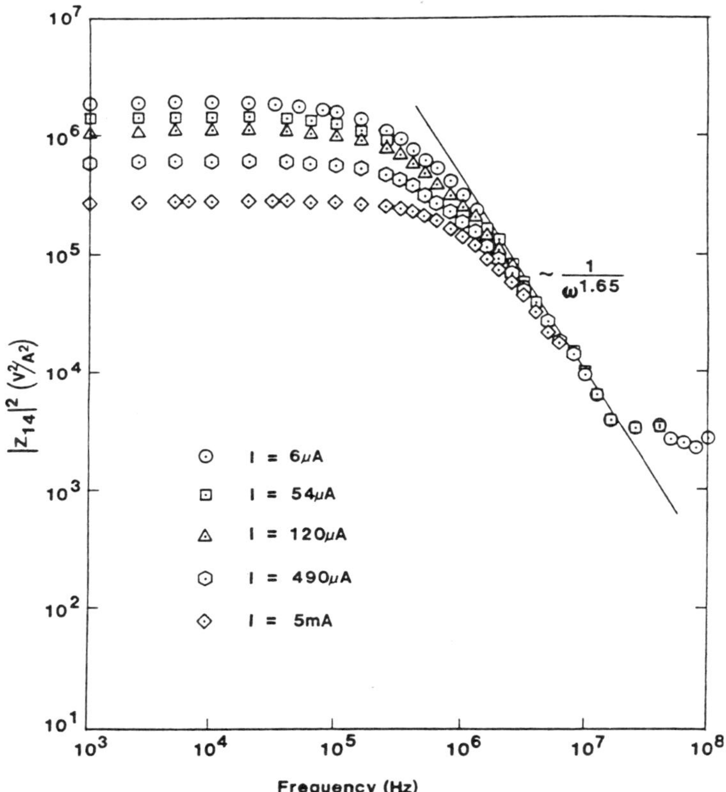


Fig. 5. $Z_{14}(f)^2$ characteristic.

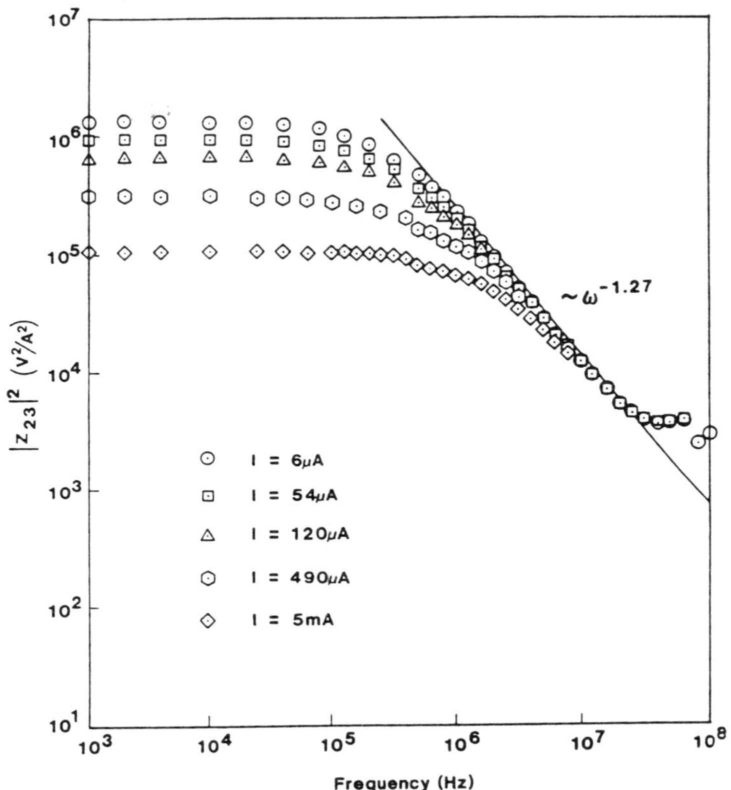


Fig. 6. $Z_{23}(f)^2$ characteristic.

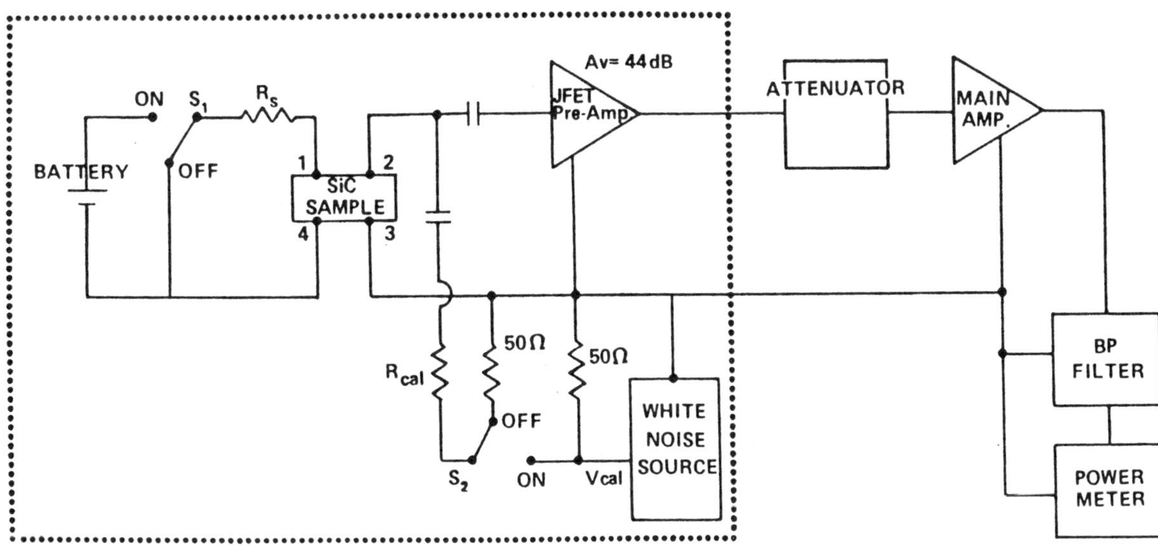


Fig. 7. Schematic of the noise measurement for the sample with four terminals and functional block diagram of the spectrum analyzer. Dotted line represents the shielding case made of steel.

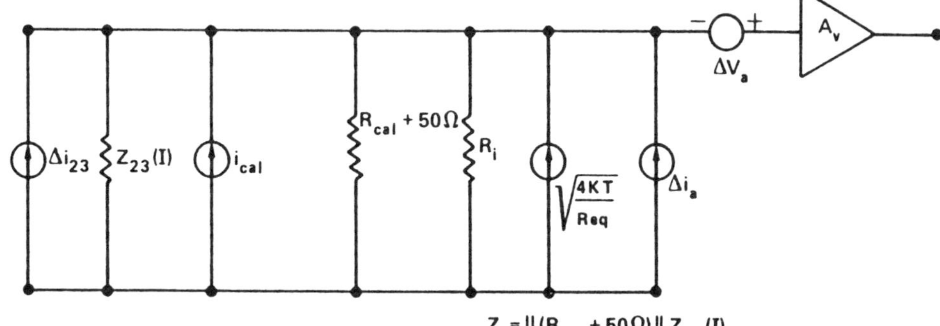


Fig. 8. Equivalent noise circuit at the input of measurement system. $Z_{23}(I)$ represents the impedance measured between terminals 2 and 3 in the presence of DC-current I. $R_{eq}$ represents the equivalent resistor ($R_{eq} = Re[Z_i]$), by which the thermal noise at the input of amplifier is defined by $S_i = 4kT/R_{eq}$. $\Delta i_a$ and $\Delta v_a$ represent the Norton and the Thevenin noise sources of the amplifier, respectively.

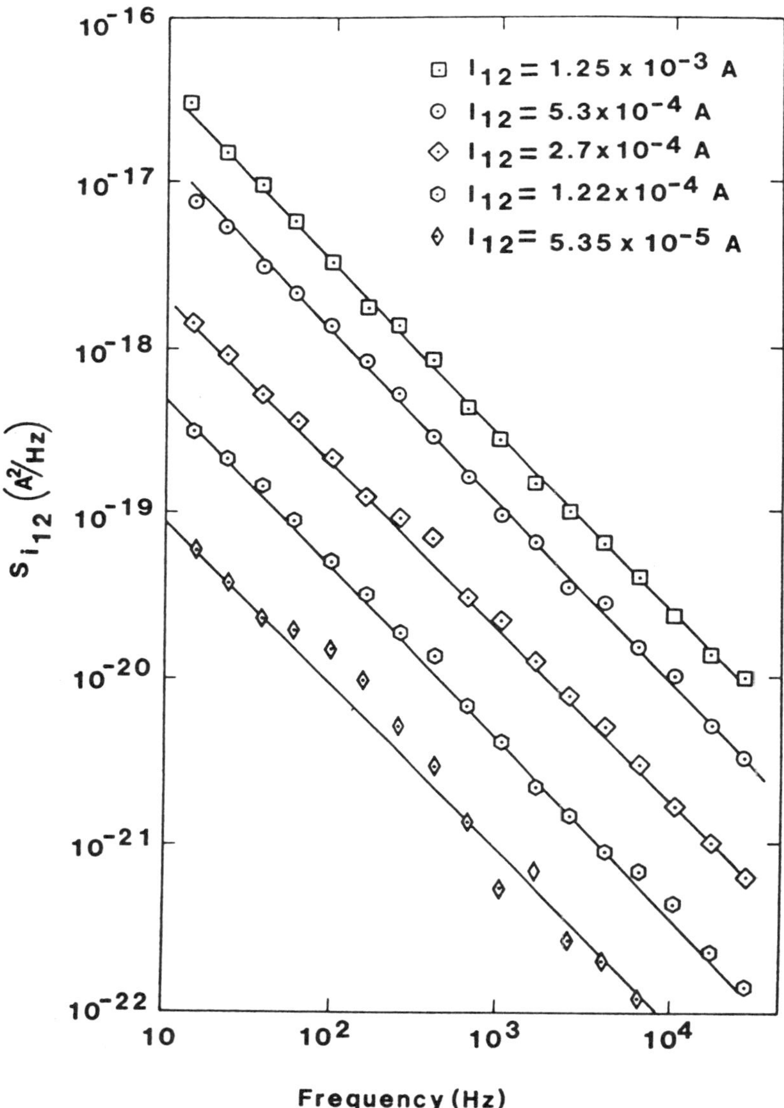


Fig. 9. Noise spectra between terminals 1 and 2 in the presence of $I_{12}$.

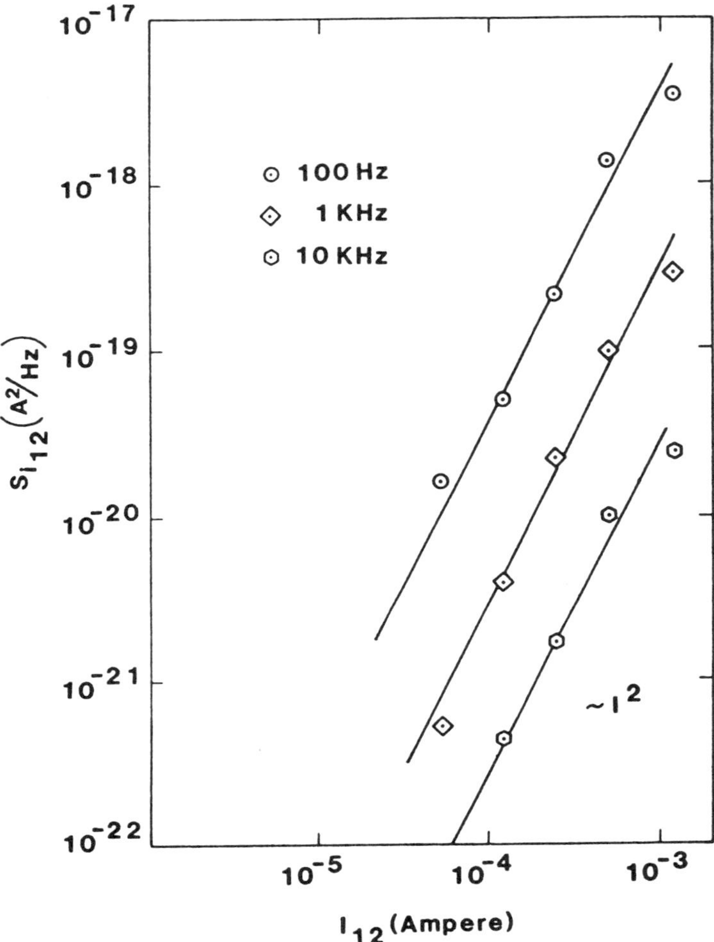


Fig. 10. $S_{i_{12}}$ at several frequencies versus $I_{12}$.

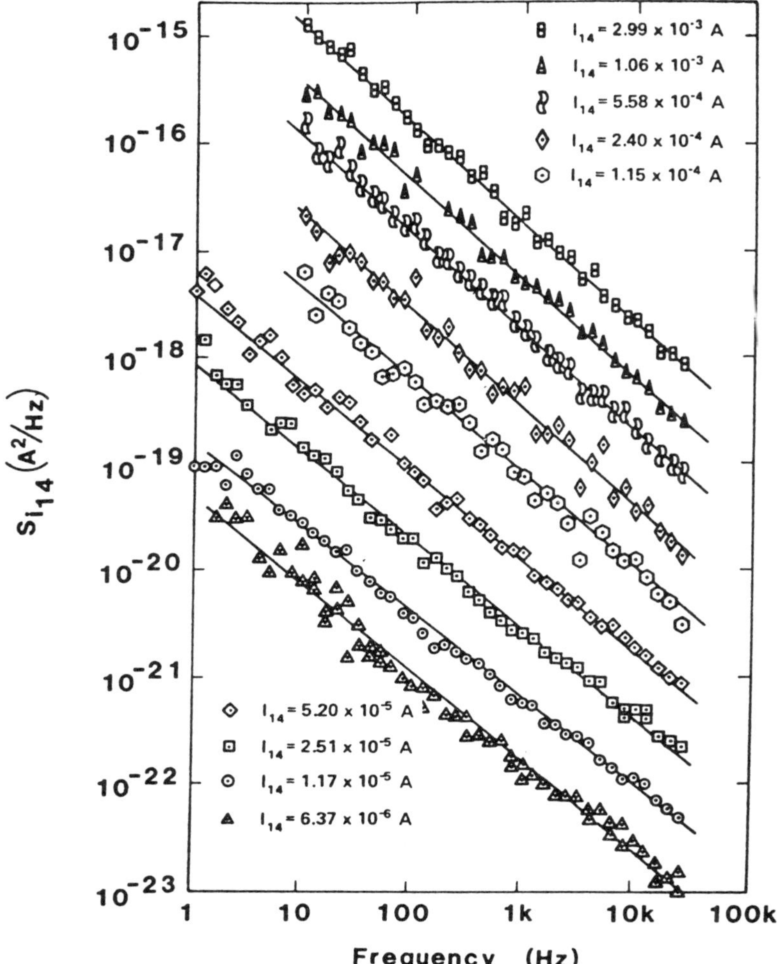


Fig. 11. Noise spectra between terminals 1 and 4 in the presence of $I_{14}$.

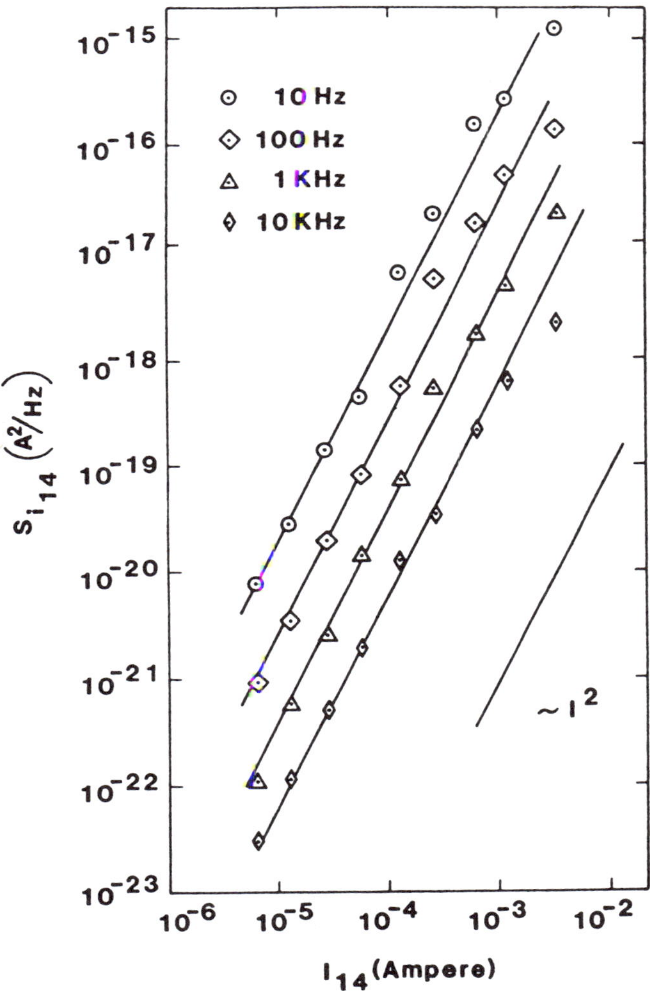


Fig. 12. $S_{i_{14}}$ at several frequencies versus $I_{14}$.

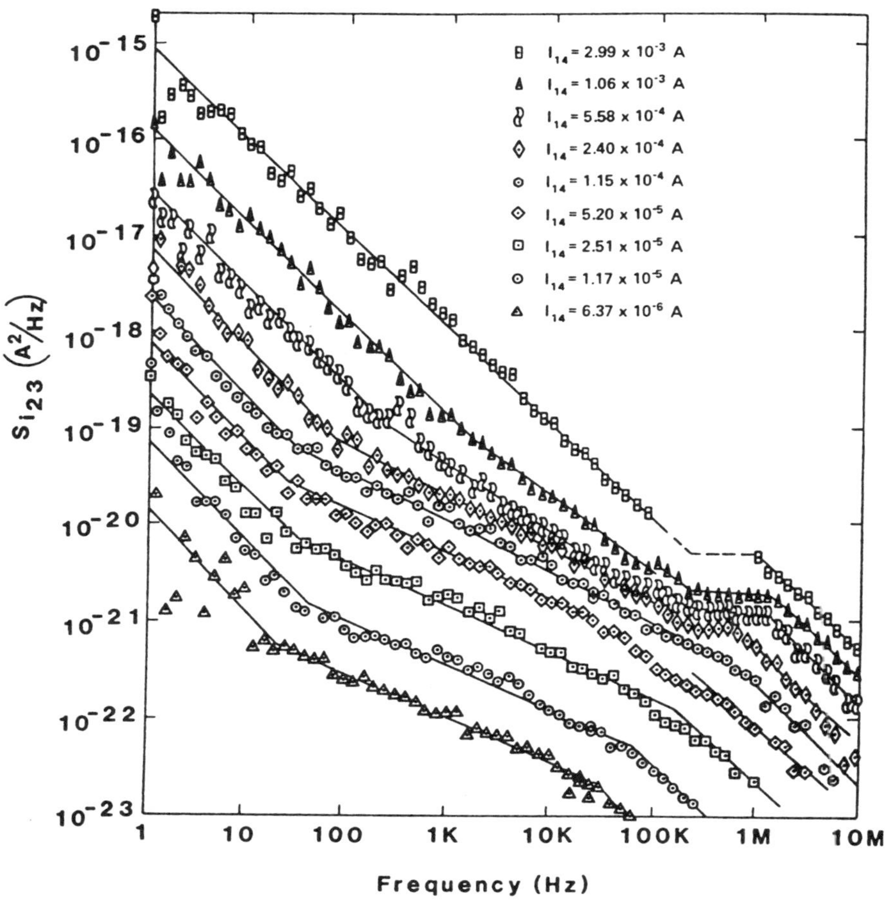


Fig. 13. Noise spectra between terminals 2 and 3 in the presence of $I_{14}$.

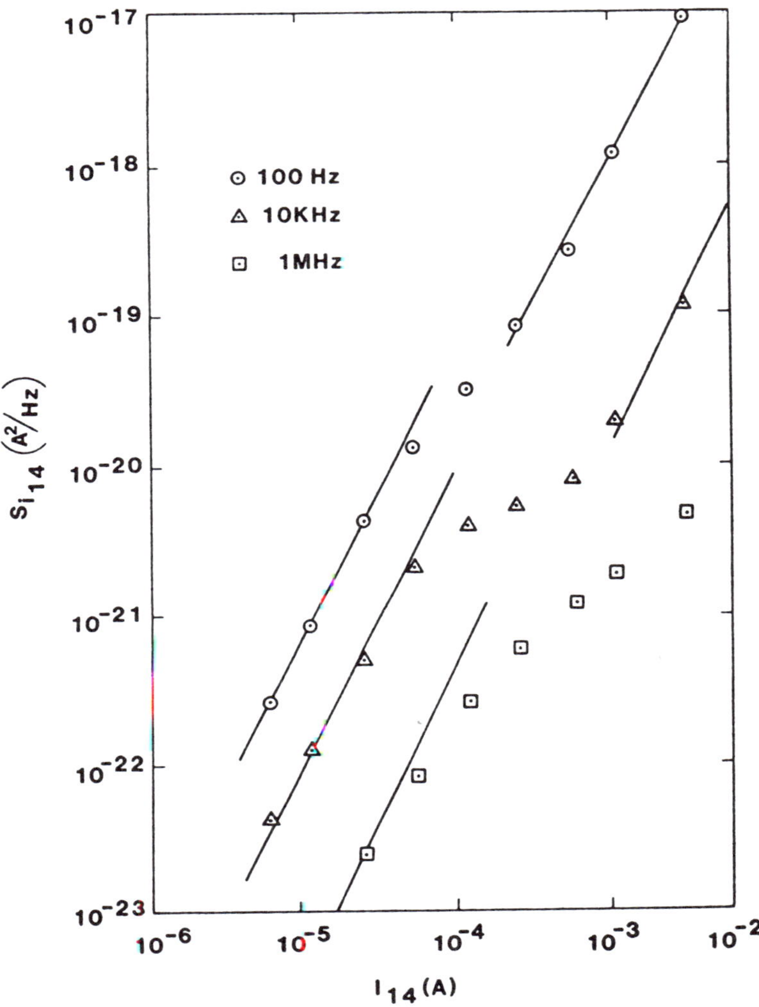


Fig. 14. Plot of $S_{i_{23}}$ versus $I_{14}$ at several frequencies.

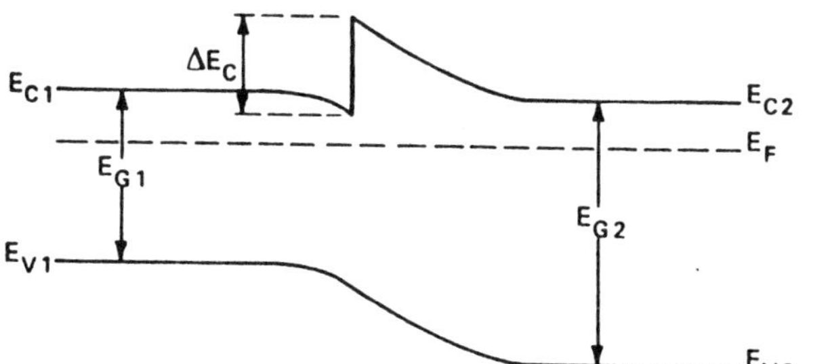


Fig. 15. Energy band structure at the polytype-polytype junction. The discontinuity in the conduction band edge is estimated by the difference in band gap between two polytypes; $E_c = E_{G_2} - E_{G_1}$. The continuity in the valence band is based on the similar property of holes between polytypes.

# LOW TEMPERATURE OXIDATION OF SiC

**S.Y. Lee and L.L. Hench**

*Department of Materials Science and Engineering*
*University of Florida*
*Gainesville, FL 32611*

## Introduction

In order for SiC to be used as structural material for high temperature applications such as heat engines or as a high band gap semiconductor for advanced electronics, it is essential to understand and control surface oxide formation. Several previous studies [1,2] have shown that minor changes in atmospheric conditions, temperature, fabrication method, densification aids, impurities, and polytypic boundaries can affect the oxidation kinetics and result in severe corrosion in some cases.

Because of these many variables affecting SiC oxidation, several models have been developed to explain the kinetics of surface oxide formation on SiC. Some emphasize the inward diffusion of oxygen through the silica scale as the rate controlling step [3,4]. Other models are based upon the desorption of CO from the SiC-$SiO_2$ interface [5]. Recent work [6-8] suggests the formation of a ternary mixed oxycarbide phase separating $SiO_2$ and SiC. Previous work in this project [9] supports the conclusion of Pampuch et al. [6,7] that an intermediate Si-O-C compound forms between an outer amorphous $SiO_2$ layer and bulk crystalline SiC. An incubation or nucleation period of several hundred minutes was required at 1100°C for the Si-O-C phase to form when dense polycrystalline SiC samples were oxidized in ambient air [9]. The intermediate oxycarbide phase was formed at the expense of specific Si-C bonds in the surface and extended for approximately 6μm underneath an 11μm oxide layer [9].

The objective of the present work is to extend the investigation of low temperature SiC oxidation to include single crystal αSiC as well as polycrystalline SiC samples. In order to improve our ability to interpret the surface reactions taking place this study is conduted in pure $O_2$. An additional objective is to compare surface reactions during continuous oxidation versus sequential oxidation. The reason for this comparison is to test the possibility that nucleation of the intermediate phase is critical for the growth of the oxide layer.

## Experimental Method

The samples studied include sintered polycrystalline SiC (General Electric, GES-004; 0.93%B and ~0.83% free C added for sintering) and small (~0.3 X 0.3 X 0.1 cm) αSiC single crystals (General Electric Company).

Oxidation was conducted isothermally at 1100°C in pure $O_2$ in an Astro Industries controlled atmosphere resistance furnace. The samples given sequential oxidation were raised from room temperature to 1100°C, requiring about 3 hr, and held at 1100°C for the times indicated, then cooled to room temperature during a 3 hr period. Surface spectral measurements were made on the samples with a Nicolet Fourier Transform IR Spectrometer (FTIRRS) in a reflection mode. After the FTIRRS analyses the sequential samples were given another exposure to $O_2$ at 1100°C, and the procedure followed again to yield the cumulative oxidation times indicated. In contrast, samples given continuous oxidation were held continuously for the times indicated without any interruptions or reheating. Each sample was analyzed with a 3mm aperture on the specular reflectance stage. A total of 96 scans were made for each oxidation treatment and the spectra shown are the computer summations of the 96 scans.

Results: Si-O peaks

Figures 1 and 2 show the FTIRRS spectra of two dense polycrystalline SiC samples sequentially oxidized at 1100°C in $O_2$ for 3, 7, 12, 18, and 25 hours. There is excellent reproducibility between the two samples.

Both samples (P-A and P-B), show extensive Si-O molecular stretching vibration peaks at 1086 $cm^{-1}$ and Si-O molecular rocking vibration peaks at 470 $cm^{-1}$ within just 3 hours oxidation in $O_2$. These peaks are much more fully-developed than those observed for air oxidation within the same time period [ 9 ]. There is a systematic shift of both Si-O peaks towards higher wavenumbers as oxidation progresses (Tables 1 and 2).

Continuous oxidation (sample P-C) at 1100°C in air also produces extensive surface oxide and Si-O peak formation within 3 hours (Figure 3). However, as shown in Table 1, the Si-O rocking peak changes much more slowly with time than was observed for the samples sequentially oxidized. The difference between continuous and sequential oxidation is even more striking when the time dependence of the location of the Si-O stretching peak is compared (Table 2). The peak position for sequentially oxidized samples shifts upwards by 8$cm^{-1}$ whereas the net change for continuous oxidation is only 2$cm^{-1}$.

Sequential oxidation of the SiC single crystal somewhat similar changes in the Si-O rocking peak as is observed for the polycrystalline sample (Figure 4 and Table 1). The peak shift upward is just a little less. However, the Si-O stretching peak shift is considerably greater for the single crystal sample, Table 2.

The time dependent changes in the intensity of the Si-O rocking peak maxima (466-478$cm^{-1}$) and the Si-O stretching peak maxima (1086-1094$cm^{-1}$) are shown in Figures 5 and 6. Figures 5A, B, and 6A, B are for sequentially oxidized polycrystalline SiC and Figures 5C and 6C are for continuously oxidized polycrystalline samples (denoted as samples H, I, J, K, L). Figures 5D and 6D show results for the sequentially oxidized SiC single crystal.

Comparison of all the samples in Figures 5 and 6 show that there is little marked difference in the time dependence of peak intensity. The abnormal change in the single crystal curves may be due to an isolated experimental variation. With the exception of that one point, a simple parabolic change in the Si-O peaks seem to be the best fit for all the samples studied.

Si-C peak:

For polycrystalline SiC samples the primary Si-C molecular stretching vibration peak is located in the range of 808 to 812 $cm^{-1}$. Figures 1, 2, and 3 again show excellent reproducibility between the samples studied. The changes in this peak during oxidation are generally the same for continuous or sequential oxidation. Figures 7A, B, and C show that the change in intensity of the peak is generally linear for both types of oxidation. For the same time of oxidation the intensity of the continuously oxidized samples is higher. This indicates a somewhat slower reaction during continuous oxidation than with the sequential exposure and its thermal cycles. There is no measureable shift in peak location for any of the samples as a function of type of oxidation or time of oxidation.

It is quite clear from Figures 1, 2, and 3 that the spectral changes in the Si-C vibrations during oxidation in $O_2$ are entirely at the expense of the vibration modes associated with the higher wavenumber of the Si-C peak. The peak intensity at 700 $cm^{-1}$ is hardly changed whereas the peak intensity at 900$cm^{-1}$ is only one-third of its pre-oxidation value. This change is more intense than that previously observed in air oxidation experiments [8,9].

The changes in the Si-C peak of the sequentially oxidized SiC peak of the sequentially oxidized SiC single crystal (Figure 4) are dramatically different from those observed in the polycrystalline samples (Figures 1-3). There are very many fewer vibrational modes attacked in the single crystal sample. Again, no shift in the peak location is observed and if one assumes that the 12 hour oxidation is an experimental artifact, as mentioned above, there is a continual nearly linear decrease in peak height (Figure 7D).

Si-O-C peaks:

Figures 1-4 show several manifestations of the intermediate Si-O-C phase. After an incubation time varying from 0 to 12 hours depending on sample type (Figure 8), a new peak appears around $964cm^{-1}$. Both sequentially oxidized polycrystalline samples show the peak well developed between 7 to 12 hours (Figures 1, 2, and 7A, B). Continuous oxidation leads to a longer incubation time of 12-18 hours. The intensity of the Si-O-C peak for the continuously oxidized sample is also less than for sequential oxidation indicating slower growth, as observed above for the other phases.

Simultaneous with the development of the $964cm^{-1}$ peak is a small peak at $1204cm^{-1}$. Figure 9 shows that this peak behaves similarly to the $964cm^{-1}$ with an incubation period followed by linear growth.

The peak positions for the Si-O-C peak developed by the SiC single crystal are different from those of the polycrystalline samples. The single crystal sample shows a broad peak ranging from 940 to $960cm^{-1}$. In order to understand the difference in peak changes of the single crystal compared with polycrystalline SiC will require studies of the effects of crystal orientation on nucleation of the Si-O-C phase.

A shoulder appears at approximately $1060cm^{-1}$ on the Si-O-Si peak also as the Si-O-C phase forms. It occurs after the incubation period and sharpens as oxidation progresses (Figures 1-4). This feature is most noticeable for the oxidized SiC single crystal (Figure 4) but can be seen clearly after 25 hours sequential oxidation as well (Figures 1 and 2). The location of this shoulder continuously decreases in wavenumber for all samples as oxidation progresses (Table 3). Thus this spectral feature also appears to be due to the nucleation of the Si-O-C phase and structural changes of this phase as oxidation progresses.

References


1. P. J. Jorgensen, M. E. Wadsworth and I. B. Cutler, "Effects of Water on Oxidation of Silicon Carbide," J. Am. Ceram. Soc., 44(6), 258-61 (1961).

2. S. C. Sinphol, "Oxidation Kinetics of Hot-Pressed Silicon Carbide," J. Mater. Sci., 11(7), 1246-53 (1976).

3. K. Motzfeldt, "On the Rates of Oxidation of Silicon and Silicon Carbide," J. Mater, Sci., 11(7), 1230-45 (1976).

4. J. A. Costello and R. E. Tressler, "Oxidation Kinetics of Hot-Pressed and Sintered α-SiC," J. Am. Ceram. Soc., 64(6), 327-31 (1981).

5. E. Fitzer and R. Ebi; pp. 320-28 in Silicon Carbide 1973, R. Marshall, J. W. Faust, Jr., and C. E. Ryan, eds., University of South Carolina Press, Columbia, SC, 1974.

6. R. Pampuch, W. S. Ptak, S. Jonas and J. Stoch, Proceed. 4th CIMTEC, St. Vincent, May 1974, P. Vincenzini, Ed., Elsevier Publ. Co., Amsterdam (1980).

7. V. A. Lavrenko, S. Jones and R. Pampuch, "Petrographic and X-ray Identification of Phases Formed by Oxidation of Silicon Carbide," Ceramics International, Vol. 7, No. 2, 1981.

8. L. L. Hench, F. Ohuchi, S. W. Freiman, C. C. Wu and K. R. McKinney, "Infrared Reflection Analyses of $Si_3N_4$ Oxidation," Proceedings of 2nd and 3rd Annual Conference on Composites and Advanced Materials, Amer. Ceram. Soc., Columbus, OH, pp. 318-330, 1980.

9. Bulent O. Yavuz and L. L. Hench, "Low Temperature Oxidation of SiC," in Ceramic Science and Engineering Proceedings, 3[9-10], 596-600 (1982).

Table 1

| T(HR) | P-A | P-B | P-C | SINGLE |
|---|---|---|---|---|
| 3 | 470 | 468 | 468 | 466 |
| 7 | 468 | 468 | 470 | 468 |
| 12 | 472 | 470 | 468 | 470 |
| 18 | 474 | 474 | 469 | 472 |
| 25 | 476 | 478 | 477 | 472 |

Table 2

Oxide Peak Shifts

| T(HR) | P-A | P-B | P-C | SINGLE |
|---|---|---|---|---|
| 3 | 1086 | 1086 | 1086 | 1088 |
| 7 | 1088 | 1088 | 1086 | 1090 |
| 12 | 1090 | 1090 | 1089 | 1092 |
| 18 | 1092 | 1096 | 1089 | 1098 |
| 25 | 1094 | 1094 | 1088 | 1100 |

Table 3

Peak Shift of the Shoulder

| T(HR) | P-A | P-B | P-C | SINGLE |
|---|---|---|---|---|
| 3 | 1086 | 1086 | 1086 | 1088 |
| 7 | 1084 | 1074 | 1086 | 1068 |
| 12 | 1068 | 1072 | 1073 | 1067 |
| 18 | 1065 | 1067 | 1072 | 1064 |
| 25 | 1063 | 1064 | 1070 | 1058 |

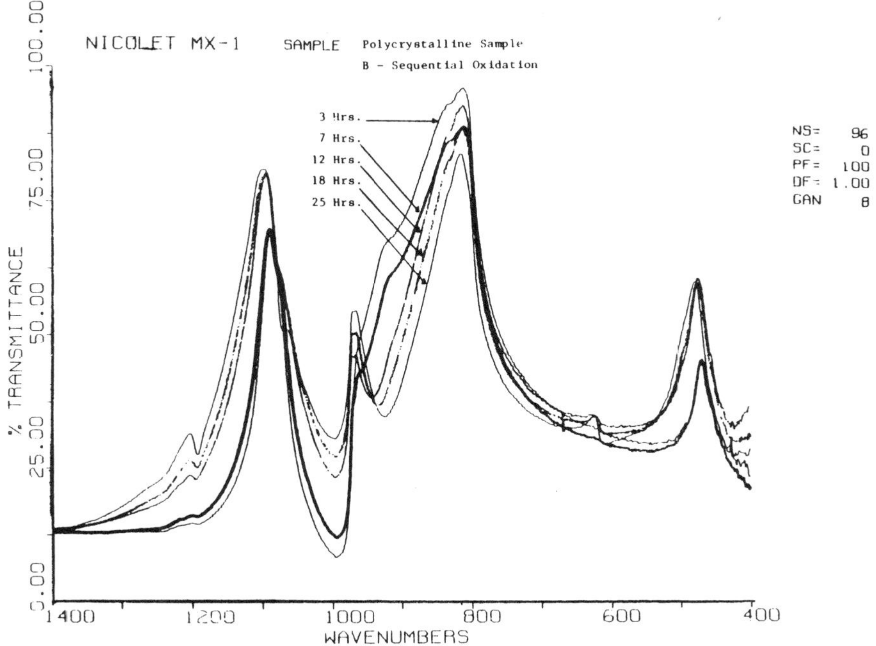


Figure 1

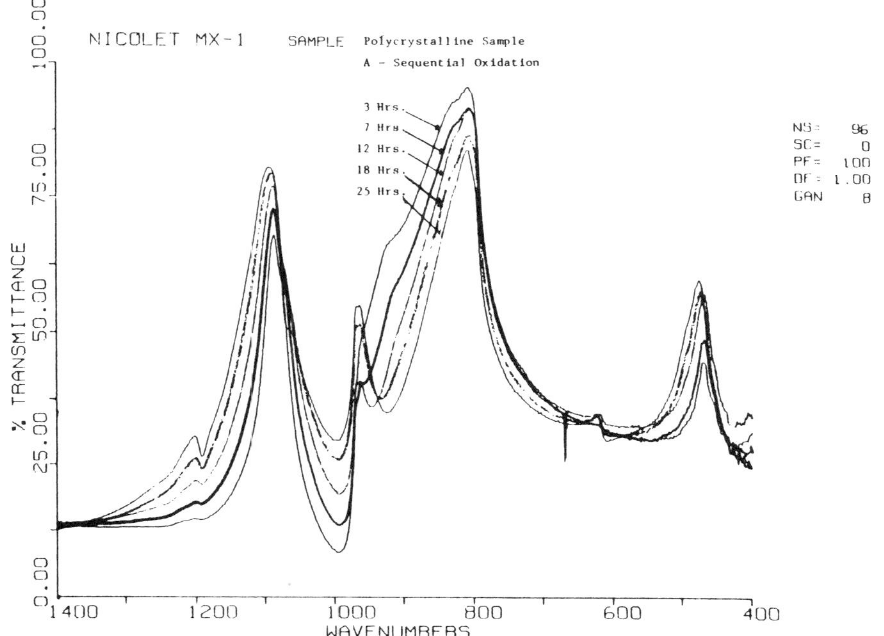


Figure 2.

NICOLET MX-1
SAMPLE Polycrystalline Sample, C,H,I,J,K,L, Continuous Oxidation
% TRANSMITTANCE
100.00
75.00
50.00
25.00
0.00
NS= 96
SC= 0
PF= 100
DF= 1.00
GAN 8
3 Hrs.
7 Hrs.
9 Hrs.
12 Hrs.
18 Hrs.
25 Hrs.
1400
1200
1000
800
600
400
WAVENUMBERS

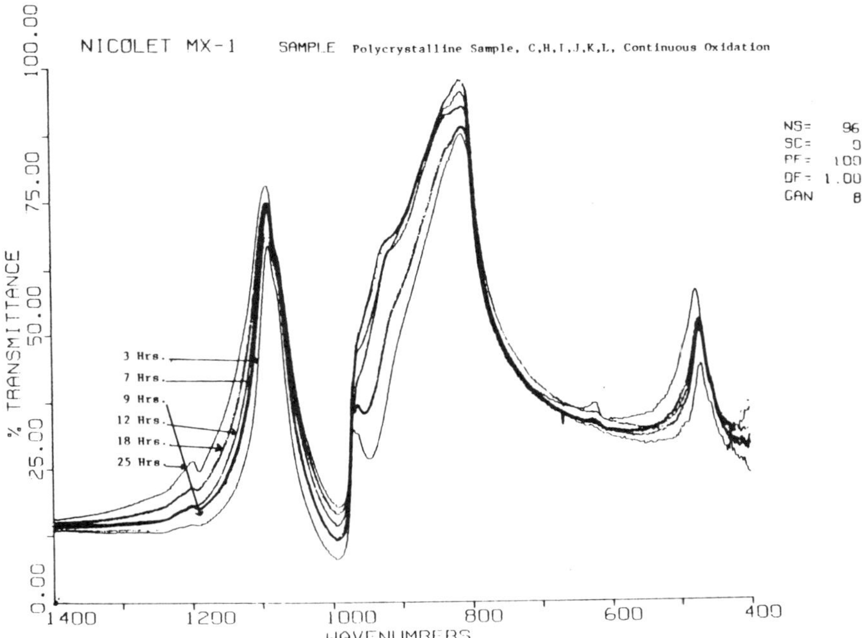

Figure 3.

NICOLET MX-1 SAMPLE Single Crystal SiC
% TRANSMITTANCE
0.00
30.00
60.00
90.00
120.00
3 Hrs.
7 Hrs.
12 Hrs.
18 Hrs.
25 Hrs.
NS= 96
SC= 0
PF= 100
DF= 1.00
GAN 8
1400
1200
1000
800
600
400
WAVENUMBERS

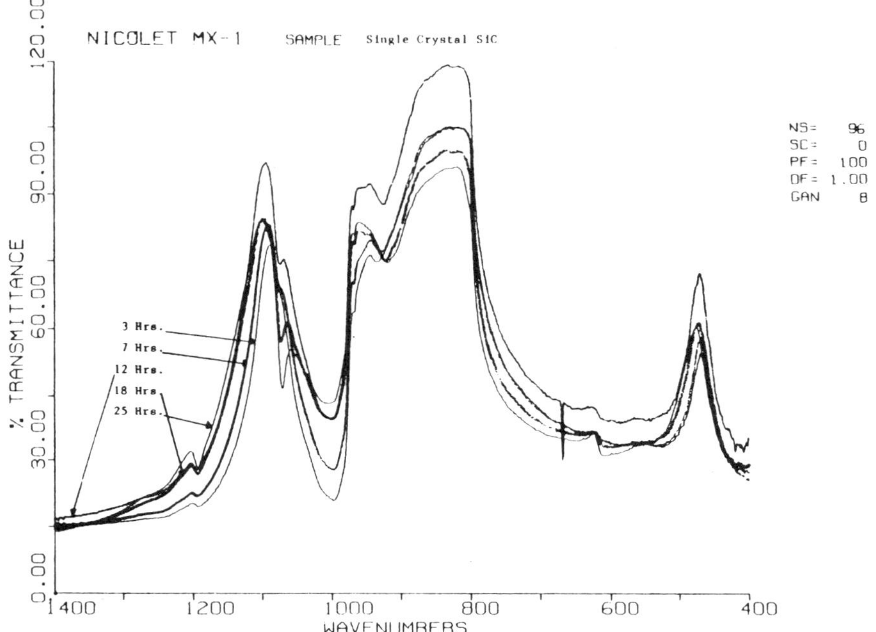

Figure 4.

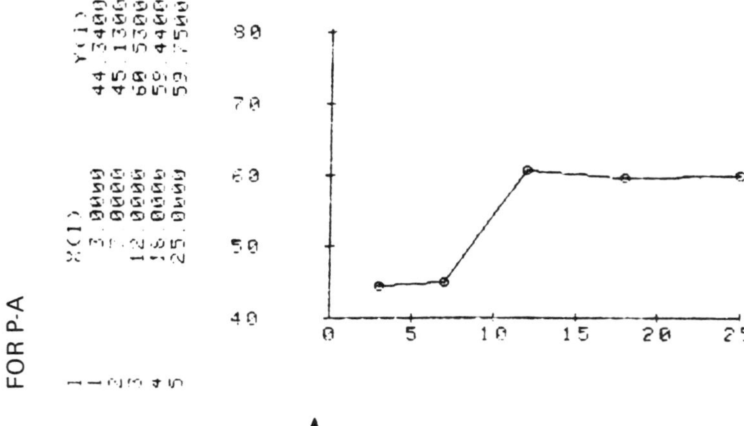


A

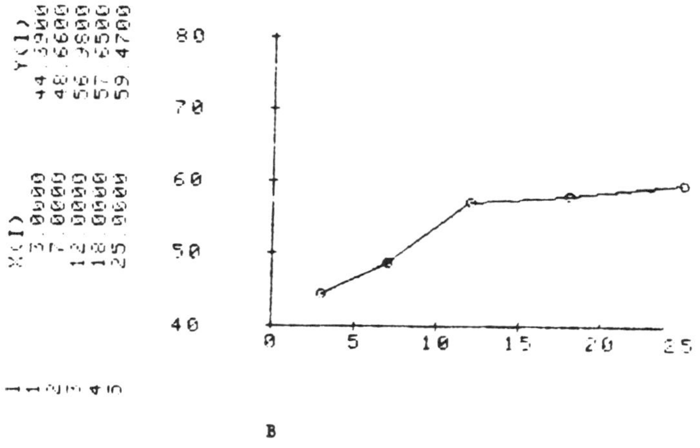


B

**Figure 5.**

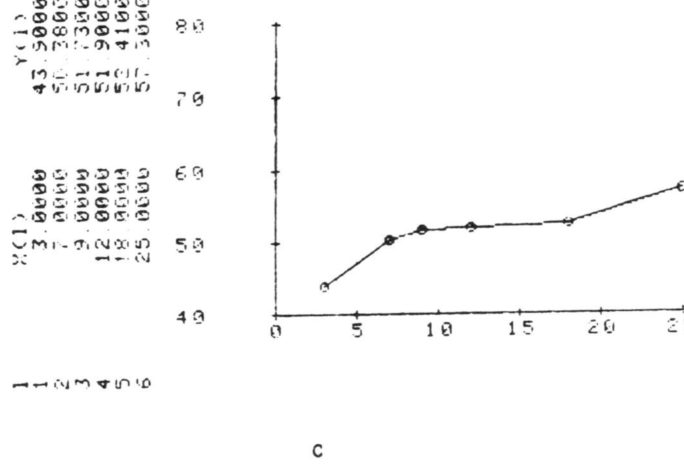


C

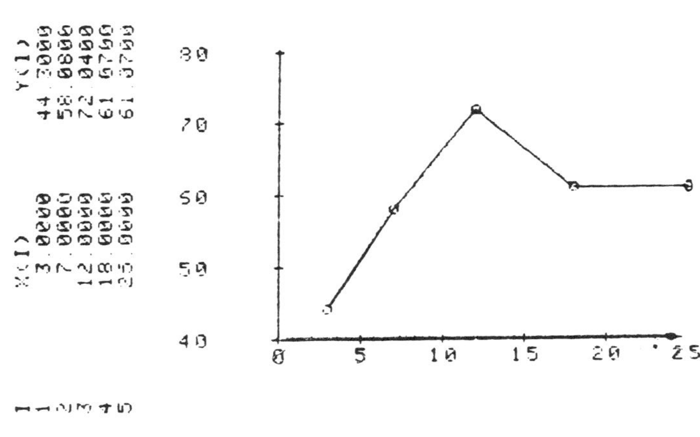


D

Figure 5. (Continued)

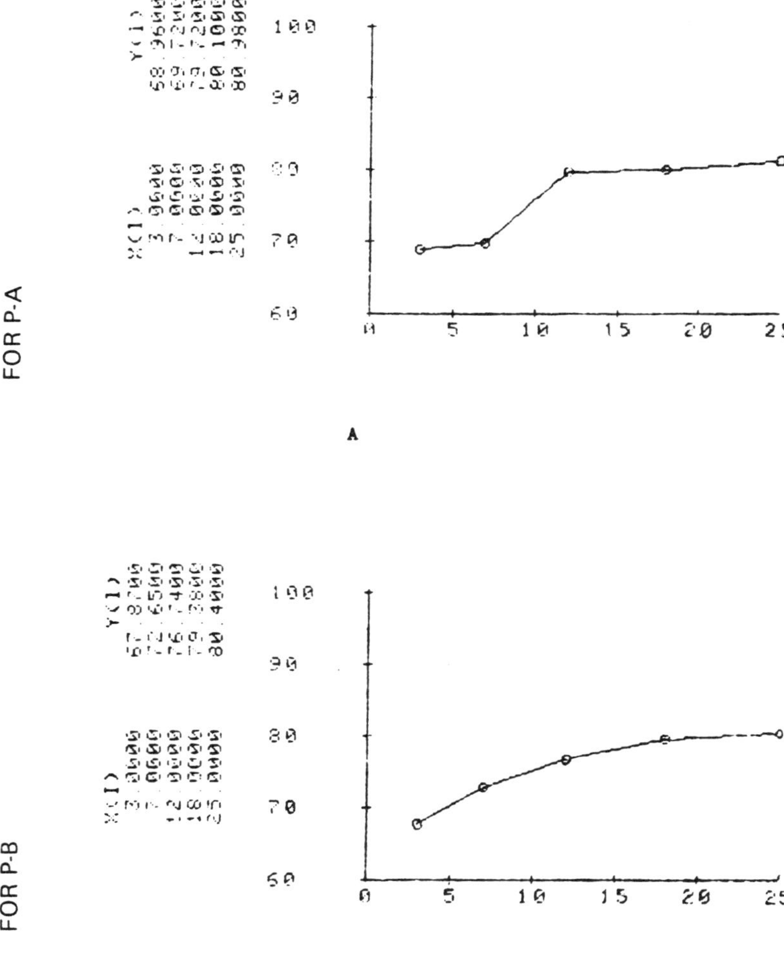


FOR P-A
100
90
80
70
60
0
5
10
15
20
25
A
FOR P-B
100
90
80
70
60
0
5
10
15
20
25
B

Figure 6.

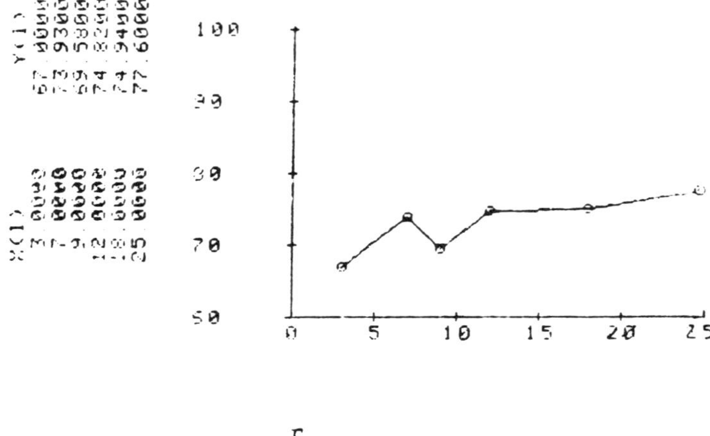


C

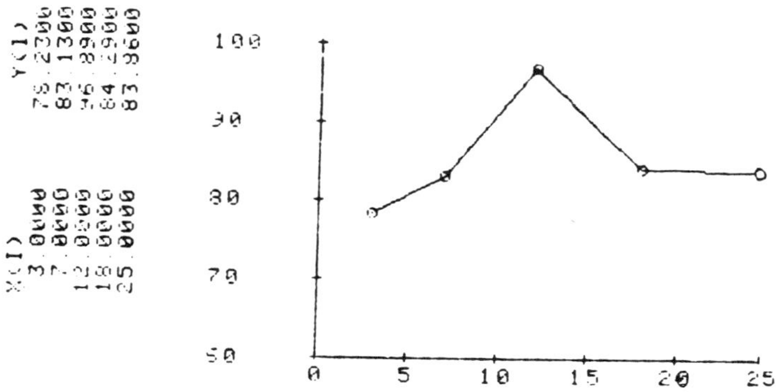


D

**Figure 6. (Continued)**

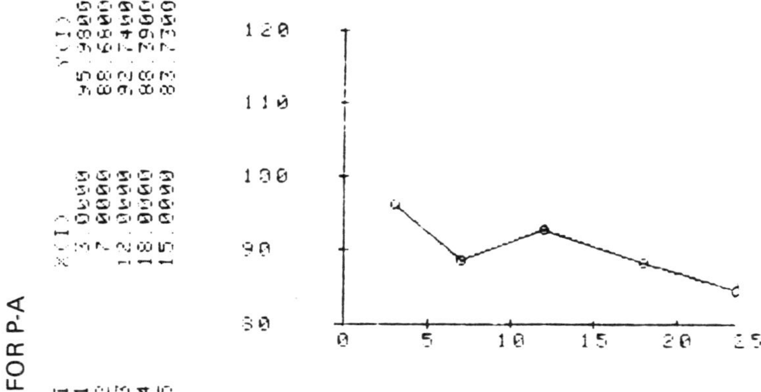

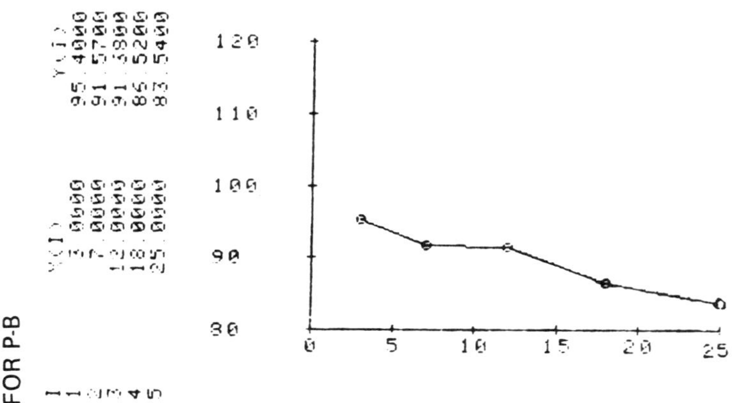


FOR P-A
120
110
100
90
80
0
5
10
15
20
25

FOR P-B
120
110
100
90
80
0
5
10
15
20
25

B

**Figure 7.**

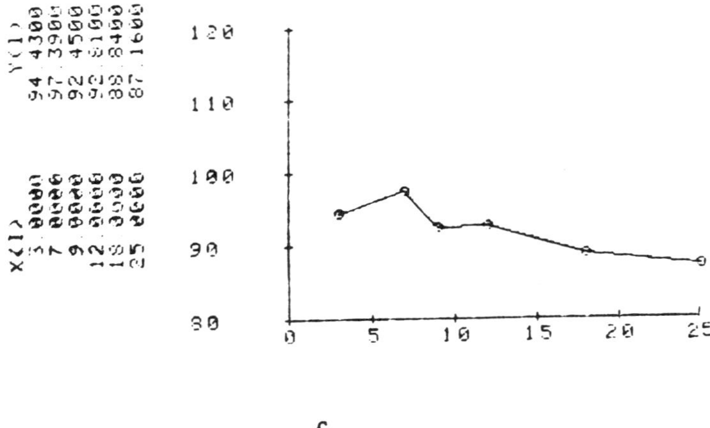


C

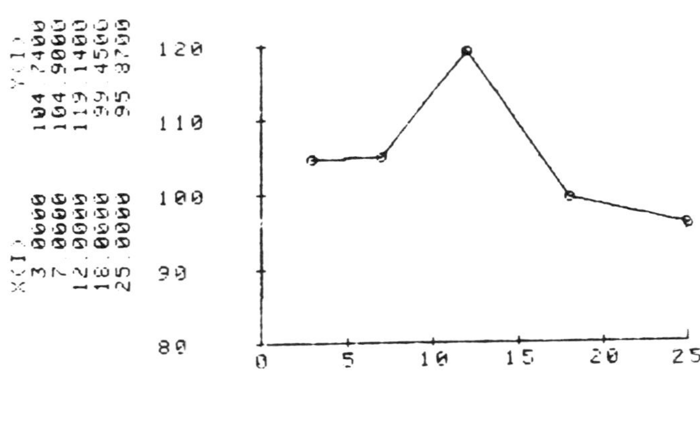


D

Figure 7 (Continued)

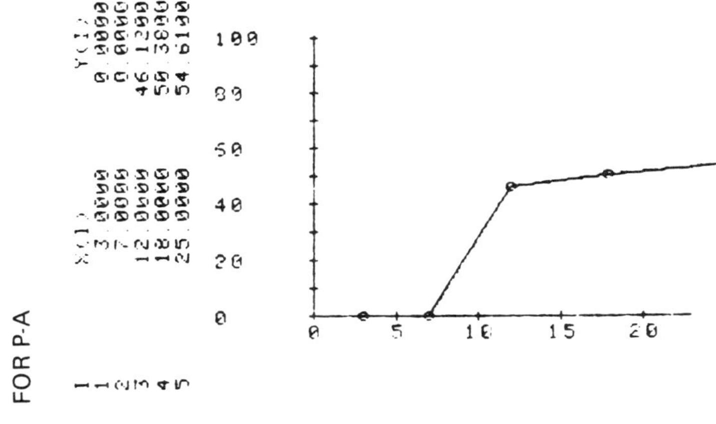


FOR P-A
I X(I) Y(I)
1 3.0000 0.0000
2 7.0000 0.0000
3 12.0000 46.1200
4 18.0000 50.3800
5 25.0000 54.6100
100
80
60
40
20
0
0 5 10 15 20

FOR P-B
I X(I) Y(I)
1 3.0000 0.0000
2 7.0000 0.0000
3 12.0000 40.7100
4 18.0000 51.3800
5 25.0000 54.7900
100
80
60
40
20
0
0 5 10 15 20

A

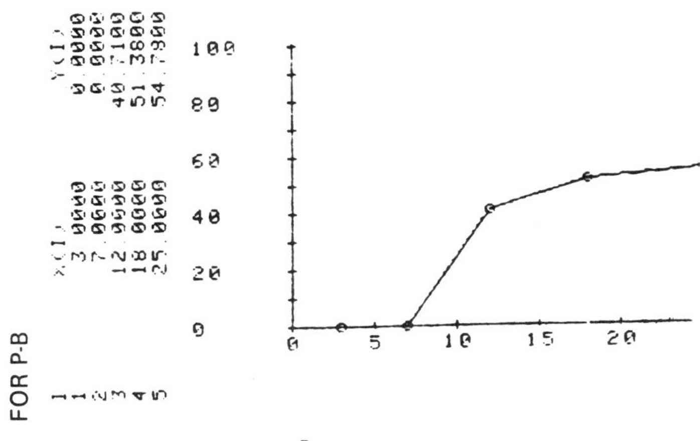


B

Figure 8.

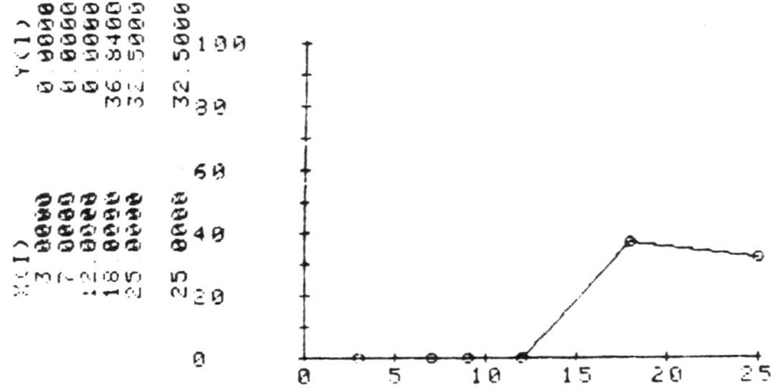


C

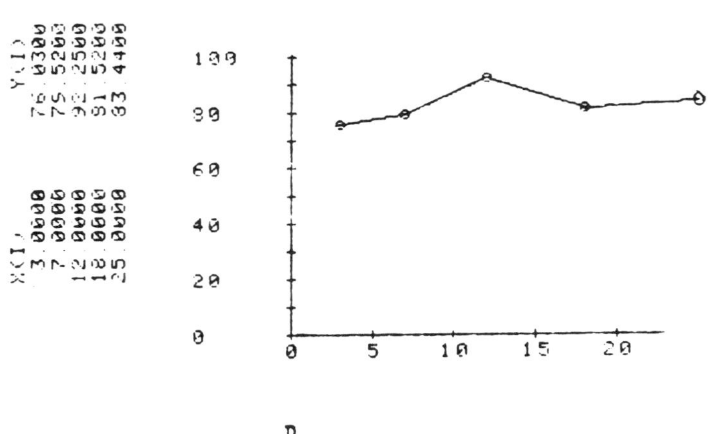


D

Figure 8. (Continued)

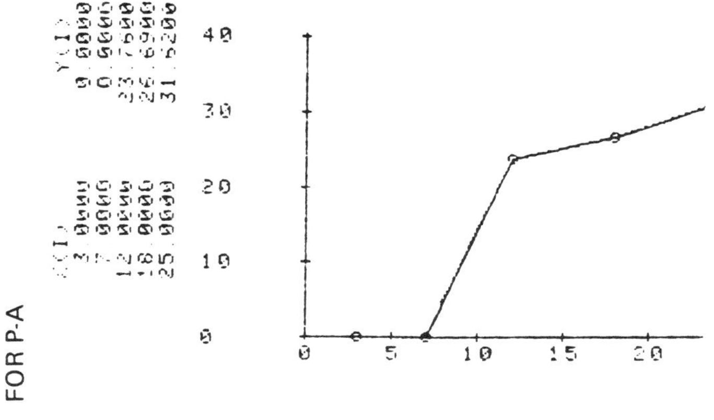


A



B

Figure 9.

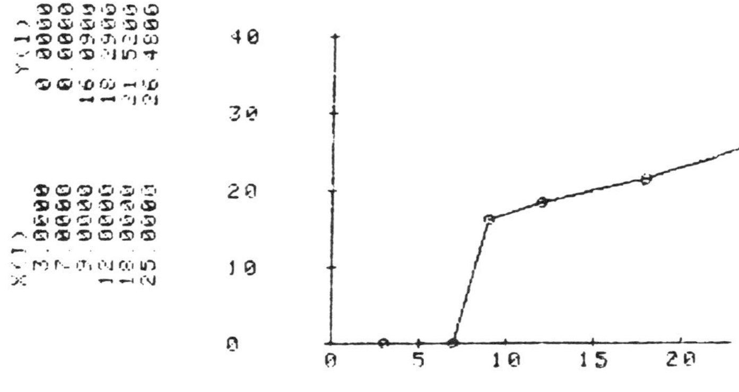


C

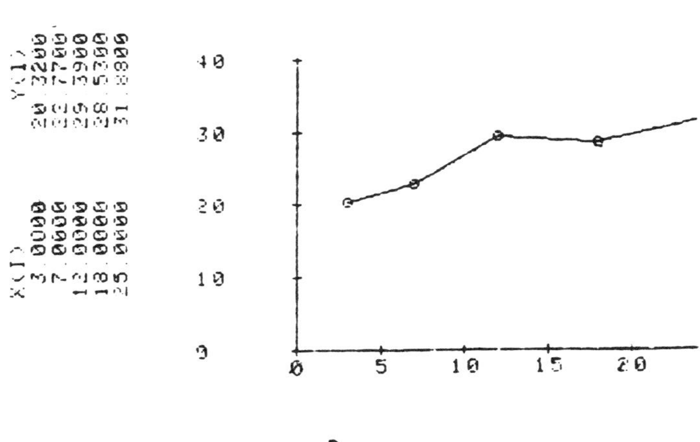


D

**Figure 9 (Continued)**

# INTERGRANULAR SEGREGATION OF BORON IN SINTERED SILICON CARBIDE

**W.D. Carter and P.H. Holloway**

*Department of Materials Science and Engineering*
*University of Florida*
*Gainesville, FL 32611*

**C. White and R. Clausing**

*Materials and Ceramics Division*
*Oak Ridge National Laboratory*
*Oak Ridge, TN 37830*

Abstract

Scanning Auger spectroscopy was used to measure Boron concentrations on in-situ fracture surfaces of oxidized sintered Silicon Carbide. Auger analysis was performed on both transgranular and small intergranular fracture areas. These data show that Boron does not segregate to or precipitate at $\alpha$-$\beta$ phase boundaries in sintered Silicon Carbide.

Silicon Carbide (SiC) is now being considered for many high-temperature structural applications, and conventional sintering is an economical method of producing these components. SiC is now commercially produced with >95% theoretical densities when boron and carbon are added to aid sintering.[1,2] The C probably reduces $SiO_2$ to SiC and CO,[2] while it has been suggested that B promotes solid state sintering by reducing the grain boundary surface energy.[3,4]

There have been many studies on the role of B and C in the sintering of SiC,[1-5] but data on the effect of these additives on other SiC properties are scarce. Past research has led to the hypotheses that B in SiC should affect such varied properties as high temperature oxidation [6,7] and the 6H→4H polytypic transformation.[4,8] To justify these hypothesized effects, the distribution of B in the SiC microstructure must be established.[4,7,9] It has been postulated that B over the solid solubility limit will segregate to the grain boundaries in SiC and accelerate sintering by increasing diffusivities along the grain boundaries.[10] This leads to the theories that excess B will appear as a segregant[7] or a B rich phase at the grain boundaries.[3] However, these theories have not been supported by direct experimental data.

One major difficulty in studying grain boundary segregation in SiC is obtaining the grain boundary as a free surface large enough for elemental analysis. Sintered SiC typically fractures transgranularly with small areas of intergranular fracture,[11] and these intergranular areas have been too small to resolve with most surface analytical instruments. However, the development of the high-resolution Scanning Auger Microprobe (SAM) has presented a method for studying compositional variations over very small areas.[12]

The purpose of this experiment was to use SAM to study segregation of B to to the grain boundaries in sintered SiC. The results of this experiment show that B does not segregate to these boundaries, but during oxidation it does appear to segregate to the oxide at the surface.

## Experimental Procedure

Sintered SiC (General Electric, GES-004; 0.93% B and ∿ 0.83% free C added for sintering)*was cut, using a Buehler diamond wafering saw, into 0.28 x 0.28 x 1.91cm billets for in-situ fracture samples or into 1.0 x 1.0 x 0.06cm slabs for angle lapping. The billets were notched 0.5cm from one end to control the fracture location. All of the samples were then successively cleaned in trichloroethylene, acetone, methanol, and ethanol agitated by ultrasonic waves. They were then rinsed in de-ionized water.

Samples were oxidized in air at 1100°C for 10, 15, or 50 hours, and at 1300°C for 10, 15, or 25 hours. The fracture billets were stored after oxidation. The oxidized slabs were angle lapped (or bevel-polished) at a shallow angle (∿ 4°). The lapping procedure was to mount the slabs on angled pistons using black-wax, and then insert the piston into a polishing collar and polish with 600 grit SiC paper, 6μm and 1μm diamond paste. The lapped samples were then cleaned as previously described.

The fracture specimens were mounted on a fracture carousel, inserted into the Perkin Elmer PHI-590 SAM, and a 12 hour bake to 200°C was used to create a vacuum of $\simeq 10^{-11}$ Torr. The specimens were then fractured and positioned for Auger analysis. The high resolution of the SAM (200nm beam diameter) allowed various composition profiles to be obtained from intergranular α and transgranular β fracture areas within

* All concentrations in this paper are expressed as atomic percent.

a fracture sample. A secondary electron detector on the SAM was used to image the fracture surface and thus position the desired areas for analysis.

The lapped specimens were mounted on a standard 60° carousel and inserted into the chamber. The system was not baked after insertion, therefore the vacuum was $\approx 10^{-9}$ Torr for analysis. For sputtering, the chamber was backfilled with research grade Argon. Sputtering was performed at 5keV-26 $\mu A/cm^2$ for 10 - 15 minutes and the chamber was then pumped down to $10^{-9}$ Torr for analysis.

The PHI-590 SAM allows analysis of very small areas, or analysis of separate spots within any area. The analyzer energy range can be multiplexed to maximize the counting times at energies of interest (e.g. data could be taken from 150-300eV for B and C peaks). These features of the PHI-590 were used extensively since the B content in the samples was near the detection limits of the system. The primary beam energies ranged from 5 to 10keV with currents of 4 or 5 nA, depending on the magnification of the scan. The general procedure was to record a 0-2keV scan on a given area and then use the multiplexer for analyzing the particular elements in the area. The data were taken in the pulse-count mode and then numerically differentiated and/or smoothed for plotting.

## Results

The equation for reducing Auger peak intensities to quantitative data is:

$$C_i = \frac{I_i/S_i}{\sum_{j=i}^{\infty} I_j/S_j} \qquad (1)$$

where $I_i$ is the measured peak to peak height (dN/dE), $S_i$ is the sensitivity factor for the $i^{th}$ element,[13] and $C_i$ is the concentration of the $i^{th}$

element in the sample. Since only Si, C, B, and O were found on these samples, the above equation can be written as:

$$C_i = \frac{I_i/S_i}{I_{Si}/S_{Si} + I_C/S_C + I_O/S_O + I_B/S_B} \tag{2}$$

To reduce the number of sensitivity factors used, it was assumed that $C_O$ and $C_B$ were small and that the equation can be reduced as follows:

$$C_C = \frac{I_C/S_C}{I_{Si}/S_{Si} + I_C/S_C} \tag{3}$$

and

$$C_B = \frac{I_B/S_B}{I_{Si}/S_{Si} + I_C/S_C} \tag{4}$$

Finally, the B concentration can be written:

$$C_B = \frac{I_B}{S_B} \cdot \frac{S_C}{I_C} \cdot \frac{I_C/S_C}{I_{Si}/S_{Si} + I_C/S_C} = \frac{I_B}{S_B} \cdot \frac{S_C}{I_C} \cdot C_C \tag{5}$$

The sensitivity factors used for initial quantification were pure element Auger intensities from published data.[13] These factors are inaccurate for SiC. Therefore, an additional correction factor was determined by assuming the average B concentration on the polished and transgranular fracture surfaces was equal to the concentration of B added before sintering (0.93% B). This correction factor was determined to be 0.81 and all the data in Tables 1 and 2 have been multiplied by this factor.

In-Situ Fracture Analysis

Grain boundary B concentrations from intergranular α-fracture areas in the SiC specimens were determined by these experiments. As part of the experiment a scanning electron microscope (SEM) was used to determine whether the α-fracture was inter- or transgranular. SiC specimens (from the same lot used in the SAM study) were fractured ex-situ and inserted into the SEM where α platelets were identified on complementary fracture surfaces. To determine whether α platelets fractured at the α-β phase boundary or transgranularly, the specimens were removed from the SEM, etched with a preferential α etch (boiling $K_3Fe(CN)_6$:NaOH, 1:1 saturated in $H_2O$) for short time intervals, and re-examined in the SEM between etchings. The result of this study showed that β-grains appeared immediately after the first etch on one of the fracture surfaces, while the matching area on the second fracture surface would slowly (through successive etches) etch away. From these results it was concluded that the SiC α platelets oriented parallel to the fracture surface were fracturing at the α-β phase boundary. It could not be determined whether the fracture surface within completely β regions were inter- or transgranular because of small grain size. However, literature data indicate that sintered SiC typically fractures transgranularly.[11]

Two specimens were fractured and anzlyzed after 15 hours at 1100°C and 1300°C respectively. A typical Auger spectrum of a fractured surface is shown in Fig. 1a. The B peak from Fig. 1a is shown at a higher sensitivity in Fig. 1b. Point analyses were performed on areas of intergranular and transgranular fracture regions on both specimens (see Fig 2). As can be seen from Fig. 1a, the fracture surfaces were very clean. The only elements present in detectable quantities are Si, C, and B. Some O was found on the 1300°C-15 h. sample; this was believed to be contamination. The B peaks in most of the scans were small and in some cases indistinguishable from the background noise (designated ND). Table 1 summarizes the B contents for intergranular and transgranular fracture areas as well as areas containing both inter- and transgranular fracture.

Table 1; Boron Concentrations on In-Situ Fracture Surfaces

| | 1100°C-15 h. | | | 1300°C-15 h. | | |
|---|---|---|---|---|---|---|
| | TG | IG | MF[2] | TG | IG | MF[2] |
| $C_B$(%) Average | 0.97 | 0.73 | 0.89 | 0.89 | 0.73 | 1.53 |
| Range | ND[1]-3.0 | ND[1]-1.9 | 0.6-1.3 | ND[1]-1.7 | ND[1]-1.9 | 0.9-2.3 |

1: ND=not detectable above noise

2: Analysis over area containing both inter- and transgranular fracture; intergranular fracture typically < 5% of area

The B concentration was mapped by monitoring the B peak intensity as the primary beam was scanned over a 2000x area (analysis time - 80 minutes). This map showed that the B was uniformly distributed over the scanned area, and that it did not form precipitates at the fracture surface.

Analysis of Polished Surfaces

Lapped samples were analyzed after 15 and 55 hours at 1100°C, and 10 and 25 hours at 1300°C. Data was taken from the polished SiC, the surface oxide, and on both sides of the surface oxide-SiC interface. The data from these samples are presented in Table 2.

Table 2; Boron Concentrations in Lapped Samples

| | 1100°C | | 1300°C | |
|---|---|---|---|---|
| | 15 h. | 55 h. | 10 h. | 25 h. |
| Oxide average | 1.1 | 0.5 | 1.4 | 0.6 |
| $C_B$ (%) range | 1.1-1.2 | 0.4-1.1 | NA[1] | 0.2-0.8 |
| SiC average | 0.9 | 0.4 | 2.6 | 1.8 |
| Range | 0.5-1.4 | 0.3-0.7 | NA[1] | NA[1] |

1: NA - data from a single scan

The data from these samples are scattered and in some cases B could not be detected above the background noise.

## Discussion

The solid solubility of B in β-SiC is ≈0.6% at 2100°C, and any B over this amount has been postulated to segregate to, or precipitate at the grain boundaries (e.g. SiC has ≈0.9% B added for sintering). However, the literature has contained speculation that B will not segregate to the grain boundaries in sintered SiC.[3,10] The results of this experiment support this view; no evidence of B segregating to the grain boundaries in sintered SiC was observed in this experiment. Further, no evidence of B-rich precipitates was found (at a resolution of 2000x) in this experiment. For example, the B concentrations in Table 1 show that the B concentration on intergranular fracture surfaces was higher but easily within the experimental error as compared to the point analysis data on transgranular regions. It was about the same or lower than for data from scanning analysis. The scanned area contained $\gtrsim$ 95% transgranular fracture area. In addition, the B concentrations are also lower than found for the polished areas shown in Table 2. Thus our conclusion is that B did not segregate to the α-β phase boundary, nor did it form precipitates to cause increased B concentrations.

The data from the SiC areas on the lapped samples correlates well with the mixed fracture data from the fracture samples. This suggests that the β-fracture was truly transgranular. The larger range of B concentrations reported for the lapped samples is attributed to the shorter data accumulation times; this causes a lower signal-to-noise ratio and results in larger fluctuations in the calculated B concentrations.

Costello et al.[6] have found that B segregates to the oxide in oxidized sintered SiC. The data from the 1100°C lapped samples support these conclusions even though fluctuations in the data were observed. The data of 1300°C do not, however these data are more limited. Castello et. al.[6] did conclude that less segregation occurred at higher temperatures.

## Conclusions

Fractured and oxidized sintered SiC samples were analyzed with a scanning Auger microscope. The results of this experiment show that Boron does not segregate to or precipitate at the $\alpha$-$\beta$ phase boundaries in sintered SiC.

## Acknowledgements

The authors wish to thank the Air Force Office of Scientific Research (grant number F49620-80-C0047), and the Oak Ridge National Laboratory SHARE program for support of this study.

Special thanks are extende to B. Yavuz (Penn. State University), and R. Padgett (ORNL) for their valuable help in this research.

## References


1. Prochazka, S., "Ceramics for High-Performance Applications", Proceedings of the 2nd Army Materials Technology Conference, Hyannis, Mass., Nov. 13-16, 1973, pp. 239-252.
2. Giddings, R.A., C.A. Johnson, S. Prochazka, R.J. Charles, "Fabrication and Properties of Sintered Silicon Carbide", General Electric Technical Information Series # 75CRD060, Schenectady, NY, April 1975.
3. Prochazka, S., R.M. Scanlan, "Effect of Boron and Carbon on Sintering of SiC", J. American Ceramic Soc., Vol. 58, No. 1-2, p. 72.
4. Bird, J.M., J.V. Biggers, "The Role of Grain Boundaries in Hot-Pressing Silicon Carbide", J. Applied Physics, Vol. 47, No. 12, Dec. 1976, pp. 5171-5174.
5. Prochazka, S., in Silicon Carbide 1973 ed. by R.C. Marshall, J.W. Faust Jr., C.E. Ryan, (Univ. of S. Carolina Press, Colombia, S.C., 1974), pp. 394-402.
6. Costello, J.A., R.E. Tressler, "Oxidation Kinetics of Hot-Pressed and Sintered $\alpha$-SiC", J. American Ceramic Soc., Vol. 64, No. 6 June 1981, pp. 327-331.
7. Costello, J.A., R.E. Tressler, I.S.T. Tsong, "Boron Redistribution in Sintered $\alpha$-SiC During Thermal Oxidation", J. American Ceramic Soc., Vol. 64, No. 6, June 1981, pp. 332-335.
8. Jepps, N.W., T.F. Page, "The 6H→3C 'Reverse' Transformation in Silicon Carbide Compacts", J. American Ceramic Soc., Vol. 64, No. 12, Dec. 1981, p. C-177.

References (continued)

9. Ogbuji, L.U., from Ultrastructure Processing and Environmental Stability of Advanced Structural and Electronic Materials, 1st annual report to A.F.O.S.R. (Bolling Air Force Base, D.C. 20032), grant No. F49620-80-C0047, Mar. 1980 to Feb. 1981, pp. 93-107.

10. Prochazka, S., C.A. Johnson, R.A. Giddings, "Atmospheric Effects in Sintering of Silicon Carbide", General Electric Technical Information Series, # 78CRD192, Schenectady, NY, October 1978.

11. Freiman, S.W., A. Williams, J.J. Mecholsky, R.W. Rice, from Ceramic Microstructures '76, ed. by R.M. Fulrath, J.A. Pask (Westview Press Inc., Boulder, Colo., 1977), pp. 824-834.

12. Czanderna, A.W., ed. Methods of Surface Analysis, Elsevier Scientific Publishing Co., Amsterdam, The Netherlands, 1975, chpt. 5.

13. Handbook of Auger Electron Spectroscopy, ed. by P.W. Palmberg, G.E. Riach, R.E. Weber, N.C. MacDonald, Physical Electronics Industries Inc., Edina, MN, 1972.

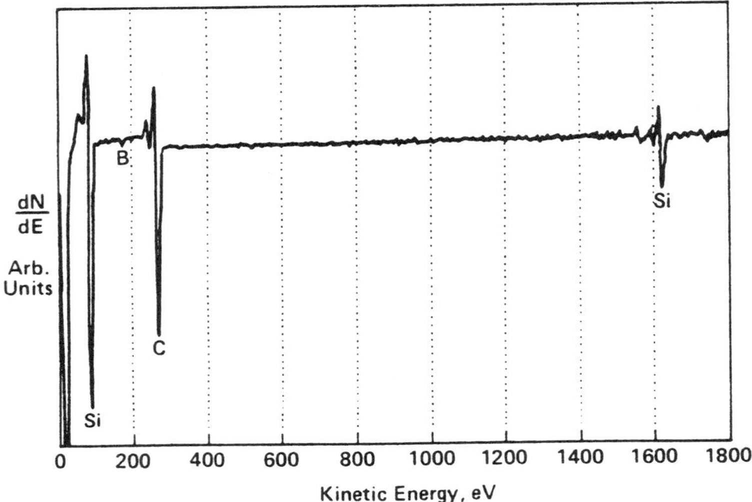


Figure 1a. Auger spectrum from the fracture surface of a SiC specimen oxidized at 1300°C for 15 hours.

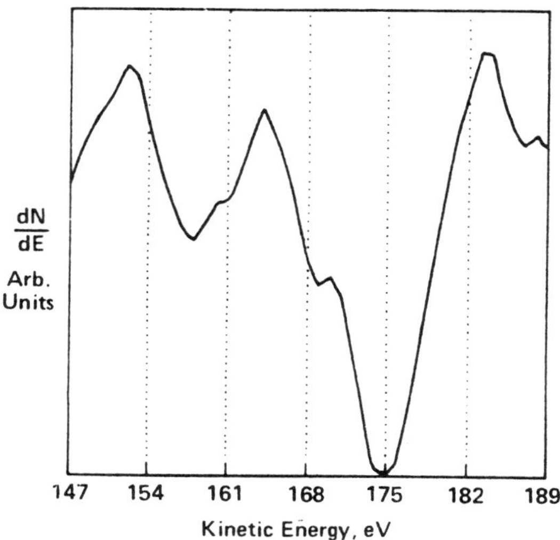


Figure 1b. The boron peak from Figure 1a at a higher sensitivity.

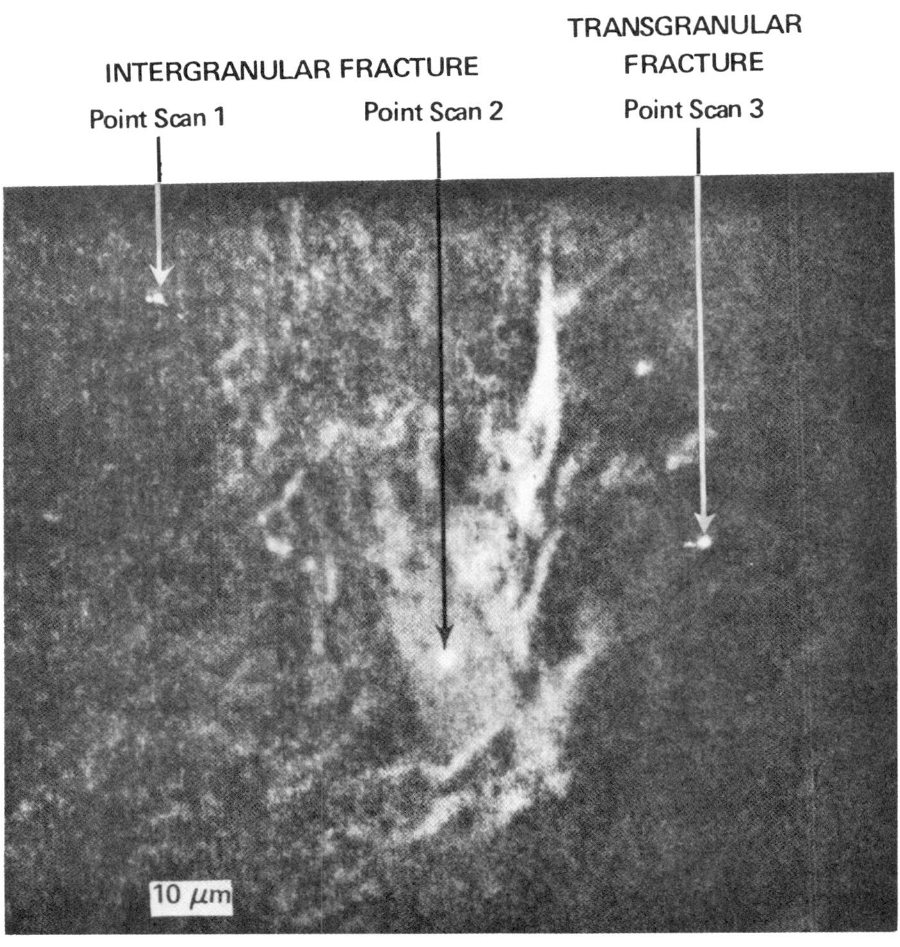


Figure 2. Photograph of the fracture surface of a SiC specimen oxidized at 1300°C for 15 hours. Points 1 and 2 are on large flat α platelets. Point 3 is on a transgranular fracture region.

# MECHANISMS OF ELECTRON STIMULATED DESORPTION FROM SODA-SILICA GLASS SURFACES

Y.X. Wang†, F. Ohuchi* and P.H. Holloway

*Department of Materials Science and Engineering*
*University of Florida, Gainesville, FL 32611*

Abstract

Electron stimulated desorption from soda-silica glass surfaces has been investigated. It was shown that $Na^+$, $O^+$, and $Si^+$ desorbed from the surface as well as other species from impure or corroded surfaces. The mechanisms of desorption include both inter-atomic and intra-atomic Auger de-excitations. For example, $Na^+$ desorbed from core-holes created either in the O level or in the Si 2p levels, while $O^+$ and $Si^+$ desorbed from core-holes created in the O 1s level. $Si^+$ desorbed only after electron bombardment created a Si-rich surface layer.

## Introduction

It is well known that the alkali Auger signal from alkali-silica glass surfaces decreases during electron bombardment.[1-6] The mechanism proposed very early to explain this phenomena was ion-electron pair formation[7] due to electron bombardment, followed by electric field-enhanced ionic transport.[5,6] Ionic transport could be increased as a result of a temperature rise due to the electron beam heating.[8] However, it has been recently shown that electron stimulated desorption (ESD)[9] may also deplete sodium at the surface.[10] A general model was developed to account for the simultaneous ionic transport and ESD.[6] This general model

---

†Visiting scientist from the Institute of Semiconductors, Chinese Academy of Sciences, Beijing, China.

*Present address: Central Research, E.I. DuPont, Wilmington, Delaware

successfully explained the incubation time observed during Auger analysis of soda-silica glass surfaces at low temperatures.[3,6] The incubation time resulted from the irradiation necessary to break the O-Na bonds and cause either ESD or ionic diffusion to occur. This model has, for the first time, put the incubation phenomena on a firm theoretical basis.

However, the mechanism of ESD of sodium from the surface has not been extensively discussed. Knotek and Fiebelman[9] have recently proposed that electron stimulated desorption from maximum-valence compounds results from inter-atomic Auger de-excitations. Preliminary data from the silica surfaces suggested that such inter-atomic Auger de-excitations also explain the $Na^+$ desorption from the glass, but the initial core-hole for the de-excitation was located on the Si, which is the second nearest neighbor of Na, rather than on a nearest neighbor.[10] The purpose of the present study is to describe the mechanisms of ESD from glass surfaces in more detail.

## Experimental

Soda-silica glass samples ($Na_2O$-$2SiO_2$) were produced by melting a mixture of reagent grade $Na_2CO_3$ and high purity $SiO_2$ at 1400°C for 24 hours in platinum crucibles. This melt was cast into graphite molds and annealed for 4 h at 250°C, or were drawn into fibers for in-situ fracture. Samples were prepared from the molded glass by cutting slabs approximately 1cm x 1cm x 2mm and diamond polished to 1μm, followed by solvent cleaning. In either case, the samples were mounted in an all metal UHV system which was evacuated to $<10^{-9}$ Torr. A PHI Model 545 scanning Auger spectrometer was used for Auger electron spectroscopy (AES) in the derivative mode (4 eVpp modulation). The coaxial electron gun was used for ESD. Secondary ions desorbed from the surface were detected by a 3M Model 610 secondary ion mass spectrometer, which included a modified UTI Model 100C quadrapole mass spectrometer. In general, the electron beam was focused to a mean diameter of ~ 20μm, then scanned over the surface. The current density was changed by varying the size of the scanned area. The beam energy and current density were varied from 50 eV to 5 keV, $10^{-8}$ to $10^{-5}$ A, and $1 \times 10^{-4}$ A/cm$^2$ to $1 \times 10^{-2}$ A/cm$^2$, respectively. In general, the electron beam struck the surface at 60° from the normal; the quadrapole was in a plane approximately 20° from the vertical and approximately 25cm from the sample.

## Results

The mass to charge ratio of desorbed positive ions from the glass fiber fractured in the vacuum is shown in Fig. 1. For data taken approximately 10 minutes after fracture at $1 \times 10^{-9}$ Torr, major species were observed at m/e = 1, 16, and 23 corresponding to $H^+$, $O^+$, and $Na^+$. However, after 90 minutes in this vacuum, positive ions representing a number of other species were observed as shown in Fig. 1b. The species at 17, 19, and 28 have increased most signifigantly and probably correspond to $OH^+$, $F^+$, and $CO^+/Si^+$ (see below). The spectrum in Fig. 1b is similar to that from glass slabs cut and polished, except that the hydrogen and fluorine signals were even higher from these surfaces.

The desorbed ion signal intensity versus electron beam bombardment time is shown in Fig. 2 for $Na^+$ and $O^+$ from a glass fiber fractured in-situ. The obvious rise in the $Na^+$ signal suggests that a conversion process is occurring, and it has been suggested previously that this process consists of trapping holes on the O lone-pair orbitals.[6] The mechanism for $Na^+$ desorption can be studied by measuring the rate of desorption versus the energy of the primary electron beam, as shown in Fig. 3. Obviously, the rate increases dramatically at approximately 100 eV, which is near the Si 2p critical energy. Thus we shall conclude that one mechanism of $Na^+$ desorption is inter-atomic Auger de-excitation resulting from a core-hole on the Si 2p level. However, the data in Fig. 3 also shows that a second increase occurs at approximately 530 eV energy, coincident with excitation of the O 1s level. Thus, another mechanism for $Na^+$ desorption is intra-atomic Auger de-excitation on the first nearest neighbor.

As shown in Fig. 1, $O^+$ desorbed from the surface along with $Na^+$. The $O^+$ signal intensity versus primary beam energy is shown in Fig. 4, along with intensity of the m/e = 28 peak. Both show a dramatic increase in signal intensity at primary energies greater than approximately 530 eV, suggesting excitation of the O 1s level causes desorption of these two species as well. However, the m/e = 28 peak wasn't observed initially from the surface as shown in Fig. 1, and its behavior versus electron bombardment time is shown in Fig. 5 where ESD data are presented for desorption from polished soda-silica glass surfaces. Desorption of $H^+$, $F^+$, $O^+$, $Na^+$, and $Si^+$ was observed from the surface. Initially, the $H^+$, $F^+$, $O^+$, and $Na^+$ desorption rates were low, went through a maximum, then decayed at times greater than ~ 30 sec. The $Si^+$ peak rose and became level at times greater than ~ 300 sec. Over this period the O Auger signal decreased as shown in Fig. 6, and the Si $L_{2,3}VV$ Auger transition developed a secondary peak at 92 eV, as shown in Fig. 7. The Auger peak at 92 eV is associated with electron beam decomposition of $SiO_2$, as discussed by several authors.[11-13] This decomposition is consistent with the decay in the $O^+$ signal from the surface at times greater than 30 sec. (Fig. 5), and with the observation of a decreased intensity for the O Auger electron peak height (Fig. 6). The decomposition of $SiO_2$ in the glass was a function of the primary beam energy and current density as ahown in Figs. 8 and 9, respectively. As the beam energy was increased from 0.8 to 1.5 keV, the ratio of 92 eV (elemental Si) to 75 (Si in $SiO_2$) eV LVV peaks decreased, indicating less electron beam decomposition occurred at higher energies. As the current density was decreased from 1 x $10^{-3}$ A/$cm^2$ to 1.6 x $10^{-4}$ A/$cm^2$, this ratio also decreased. Similar variations were observed in the intensity of the $Si^+$ peak with respect to beam energy and current density.

## Discussion

The measurement of secondary ion current as a function of beam energy (i.e. threshold measurements) suggests that the Auger de-excitation mechanism causes ESD from soda-silica glass surfaces. This probably results from a sequence of events such as shown in Fig. 10, where a primary electron incident upon the glass surface creates a Si 2p core-hole, and a de-exciting electron drops from the Si-O bonding orbitals or the O nonbonding orbitals. Such events would result in a $Si^{4+}$-$O^0$-$Na^+$ configuration, at least in a very simplified scheme. However, the rate of $Na^+$ desorption initially showed an increase followed by a decrease; the decrease would be expected from ESD, but not the initial rise (Fig. 2). The increase was postulated to result from holes being trapped on the O lone-pair orbitals.[10] As a result, the inter-atomic Auger de-excitation discussed above would leave the surface complex in a $Si^{4+}$-$O^+$-$Na^+$ configuration, and desorption of $Na^+$ and $O^+$ would be expected. Furthermore, the increase in the $Na^+$ desorption rate when the primary electron energy crossed the critical energy for excitation of the O 1s level clearly shows that not only does the inter-atomic Auger de-excitation cause $Na^+$ desorption, but intra-atomic de-excitation can also result in ESD. A mechanism similar to the inter-atomic case would be envisioned here for the intra-atomic mechanism, where a hole would be initially trapped on the O lone-pair orbitals. The trapped hole would have a long lifetime due to the presence of sodium and/or the surface. Auger de-excitation of a hole in the O 1s core level would result in an unstable positive charge configuration of the $Si^{4+}$-$O^{+}$-$Na^{+}$ surface complex which again would dissociate.

Similarly, ESD of $O^{+}$ and $Na^{+}$ would be expected when excitation occurs in the Na 1s level or the Si 1s level. However, due to a poor signal to noise ratio, and problems with rapid decay of the Na surface concentration, increases at the critical energies for these core levels were not detected. We speculate that core holes created in these levels would cause ESD, however, the desorption rates are much smaller than for the O 1s or Si 2p levels due to a lower ionization cross-section. An increase in the ESD rate was not observed when the Si 2s critical energy was exceeded. A dramatic increase was not anticipated since core holes created in the Si 2s level rapidly decay by a Coster-Kronig transition to a core hole in the 2p level. Such a transition only creates one positive charge in the Si-O bonding orbitals, and as a result the ESD would not be expected to occur as quickly. The O 2s level may cause ESD, but its critical energy was too low to measure with our equipment. Excitation of the O 2s level may explain small yields in the secondary ions observed for primary energies somewhat below 100 eV, but the energy is just barely adequate to cause desorption (an excess of only 2.2 eV).[10]

Ohuchi and Holloway[6] have reported $Na^{+}$ desorption cross sections of $1 \times 10^{-18}$ and $3.2 \times 10^{-19}$ $cm^2$ for 0.5 and 4 keV electron beam energies, respectively. The desorption cross sections for the Auger de-excitation mechanisms can be estimated. It is equal to the product of the inner shell ionization cross section (Si 2p) and the probability of ion desorption without neutralization. Maguire[14] has calculated the Si 2p ionization cross section; for primary energies of 0.5 and 4 keV, cross sections of $4 \times 10^{-18}$ and $2 \times 10^{-18}$ $cm^2$ were reported. Similar values for the cross section were obtained experimentally by Mayer and Vrakking.[15,16] Since the inter-atomic transition rate is very fast ( approximately $10^{-14}$ sec.[17]), the probability of Auger neutralization is limited only by the density of states of the glass valence band. Because of the localized electron states in this band, the probability of Auger neutralization is low.[18] The survival probability of an ion is estimated then to be between 0.01 and 0.1.[17] Thus, the estimated sodium ESD cross section is approximately $10^{-19}$ $cm^2$ or less. This value is smaller than that determined by experiment, which results from the fact that we have considered only one channel for ESD of $Na^+$. The total cross section for ESD should be the summation of all the operative channels (inter-atomic and intra-atomic) and the summation would increase the theoretical cross-section.

As reported earlier, it is well known that electron beams will cause $SiO_2$ to decompose. ESD has been reported to be the mechanism by which this decomposition occurs.[11,12] However, the earlier postulates were based upon the bonding orbital interaction model of ESD postulated earlier by Menzel and Gomer,[19] and Redhead.[20] Ohuchi and Holloway[6,10] first postulated that inter-atomic Auger de-excitation was the mechanism by which ESD occured from glass surfaces. In addition, the desorption of $Si^+$ from these surfaces has been reported in this study. We believe that the mass to charge peak at 28 a.m.u. is truly representative of $Si^+$ desorption, rather than desorption of a $CO^+$ species, since (1) it is associated with a primary beam whose energy exceeds that of the critical energy for the O 1s level, (2) it correlates well with the Auger $L_{2,3}VV$ peak shape change representative of decomposition of $SiO_2$, and (3) there was some evidence of Si isotopes in the desorbed spectra (although the signal to noise ratio was very poor for isotope determination). Desorption of $Si^+$ from the surface may well be expected, based upon the models shown in Fig. 11. In this model, ESD from the surface causes a depletion in the surface layer of both Na and O (see Fig. 6). As a result, it is anticipated that the 92 eV elemental Si signal in the Auger spectra results from Si which has a limited number of bonds to other Si atoms in the surface and which has either a single or very few bonds with O. The increase in the $Si^+$ signal for primary beam energy > 530 eV suggests that creation of a core hole in the O 1s level resulted in $Si^+$ desorption. It is postulated therefore, that Si bonded to a limited number of O atoms on the surface would experience the creation of a core hole in the O to which it was bound; inter-atomic Auger de-excitation on the O atom would result in a positive silicon ion core being next to a positive oxygen ion core. As a result, this unstable complex would cause desorption of the silicon atom from the surface. Desorption of $Si^+$ explains why a steady state is commonly observed in the decomposition of $SiO_2$ surfaces, even though an $O^+$ signal continues to be observed over long periods of time. $Si^+$ desorption provides a mechanism by which the electron beam by itself may continue to erode the glass surface.

The dependence of $SiO_2$ decomposition upon the primary beam energy and current density shown in Figs. 8 and 9 are consistent with data in the literature by Thomas[11] and Strausser and Johannessen.[12] For example, Strausser and Johannessen report decomposition for electron doses greater than $10^{19} e/cm^2$ at energies greater than 2 keV. The data in Fig. 8 show that decomposition was observed for the soda-silica glass surfaces at electron doses greater than $10^{16} e/cm^2$ for an energy of 1.4 keV. While this damaging electron dose is much less for soda-silica glass as compared to $SiO_2$, Strausser et. al[21] also have shown that electron beam damage will occur at lower doses for silicon oxides of lower stoichiometry. While the data on current density shown in Fig. 9 are consistent with Thomas' data in that decomposition became more severe as the current density was increased, the dependence upon beam energy reported by Thomas is different from those reported in the present study.

Knotek and Houston[22] have reported that $O^+$ will desorb from silicon surfaces with a low oxygen coverage. However, they report that $O^+$ does not desorb unless $H^+$was also observed to desorb from the surface. The dependence of ESD upon hydrogen being present was not investigated thoroughly in this study, however $H^+$ was always present due to the ion exchange corrosive reaction which occurred on the surface, either in the UHV system or as a result of exposure to atmosphere. The effect of hydrogen on ESD is unexplained, however the present data for ESD from soda-silica glass may help to explain this effect. As reported above, the decomposition of the glass surface occurred at lower electron doses than for a quartz surface.[12] We have laso speculated that the electron stimulated desorption occurs most rapidly when an electron hole is trapped on the oxygen lone-pair orbitals. As demonstrated by Ohuchi and Holloway,[6] the lifetime of this trapped hole is relatively long. The trapped hole causes faster rates of electron stimulated desorption (see Fig. 2), and therefore faster decomposition of the glass surface. The effect of hydrogen in $SiO_2$ may be to increase the lifetime of the hole trapped on the oxygen lone-pair orbitals, causing the desorption rates to be signifigantly higher in this case.

Finally, a positive ion with a mass to charge of 19 was observed in this study, and this species is assigned to $F^+$. Initially, it was thought this peak was representative of $H_3O^+$ desorbing from the surface. Many authors have speculated that the hydronium ion is created by corrosion of glass surfaces.[23] To attempt to answer this question, surfaces of corroded and uncorroded glasses were analyzed in the present study and the peak at m/e = 19 increased for ESD from a corroded glass surface. However, it is now apparent that F is a ubiquitous contaminant in ultra-high vacuum systems; this confuses the issue of $H_3O^+$ desorption.

## Summary

The mechanism of electron stimulated desorption from soda-silica glass surfaces involves Auger de-excitaitons, but these de-excitations can be either inter-atomic or intra-atomic Auger events. $Na^+$ can be desorbed by creating core holes in the second nearest neighbor Si atoms or on the first nearest neighbor O atoms. $Si^+$ desorbed only after the electron beam created a Si-rich surface layer. The dependence of electron stimulated desorption on beam energy and current density was discussed, as were the magnitudes of the ESD cross sections.

## Acknowledgement

This research was supported by the Air Force Office of Scientific Research, Grant # F49620-80-C-0047.

References


1. R.A. Chappel and C.T.H. Stoddard, Phys. Chem. Glasses 15, p. 130 (1974).
2. A.E. Clark, Jr., C.G. Pantano, Jr., and L.L. Hench, J. Am. Ceram. Soc. 59, 37 (1976).
3. C.G. Pantano, D.B. Dove, and G.Y. Onoda, Jr., J. Vac. Sci. Technol. 13, 414 (1976).
4. P.T. Dawson, O.S. Heavens, and A.M. Pollard, J. Phys. C11, 2183 (1978).
5. F. Ohuchi, M. Ogino, P.H. Holloway, and C.G. Pantano, Jr., Surf. Interface Anal. 2, 85 (1980).
6. F. Ohuchi and P.H. Holloway, J. Vac. Sci. Technol. 20, 863 (1982).
7. H.L. Hughes, IEEE Trans. Nucl. Sci. NS-16, 195 (1969).
8. F. Ohuchi, D.E. Clark, and L.L. Hench, J. Am. Ceram. Soc. 62, 500 (1979).
9. M.L. Knotek and P.J. Fiebelman, Phys. Rev. Lett. 40, 964 (1978).
10. F. Ohuchi, Ph.D. Thesis, University of Florida, 1981.
11. S. Thomas, J. Appl. Phys. 45, 161 (1974).
12. J.S. Johannessen, W.E. Spicer, and Y.E. Strausser, Faraday Discuss. 60, XXX (1975).
13. C.J. Pantano and T.E. Madey, Appl. Surf. Sci. 7, 115 (1981).
14. T.H. Di Stefano and D.E. Eastman, Phys. Rev. Lett. 27, 1560 (1971).
15. F. Meyer and J.J. Vrakking, Phys. Lett. 44A, 511 (1973).
16. J.J. Vrakking and F. Meyer, Surf. Sci. 47, 50 (1975).
17. H.D. Hagstrum, in Electron and Ion Spectroscopy of Solids (Plenum Press, New York, 1978), 273.
18. M. Salmeron, A.M. Barrow, and A.M. Rojo, Surf. Sci. 53, 689 (1975).
19. D. Menzel and R. Gomer, J. Chem. Phys. 41, 3311 (1964).
20. P.A. Redhead, Can. J. Phys. 42, 886 (1964).

References (continued)

21. Y.E. Strausser and J.S. Johannessen, NBS Special Publication 400-23, ARPA/MBS Workshop IV, Gaithersburg, Maryland, April 23 and 24, 1975.

22. M.L. Knotek and J.E. Houston, to be published.

23. R.H. Doremus, Glass Science (John Wiley and Sons, New York, 1973).

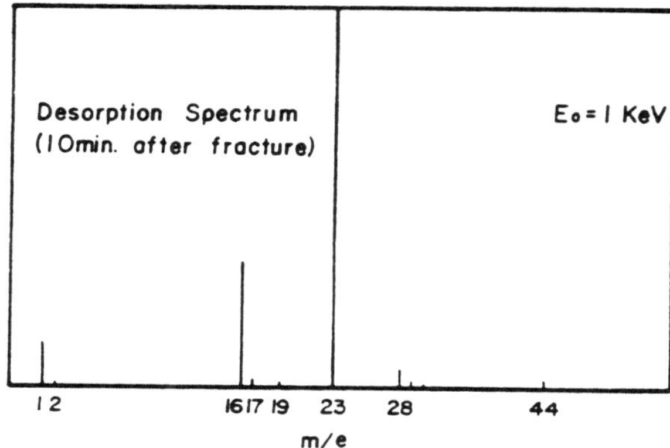


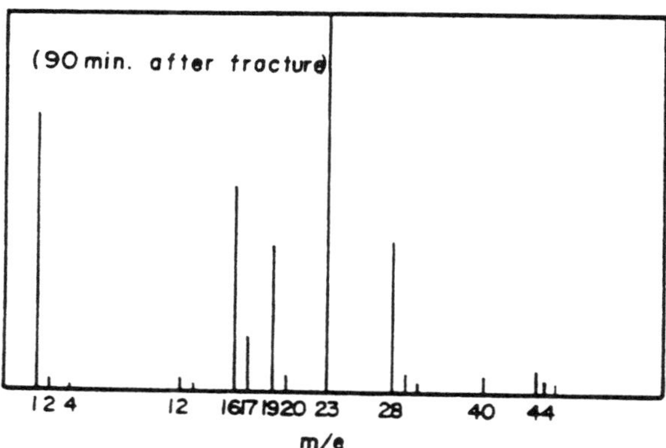


Fig. 1. ESD spectra from a soda-silica glass surface exposed to a pressure of 1 x $10^{-9}$ Torr for 10 minutes (a) and 90 minutes (b) after in-situ fracture. A 1 keV, 1 $mA/cm^2$ electron beam was used.

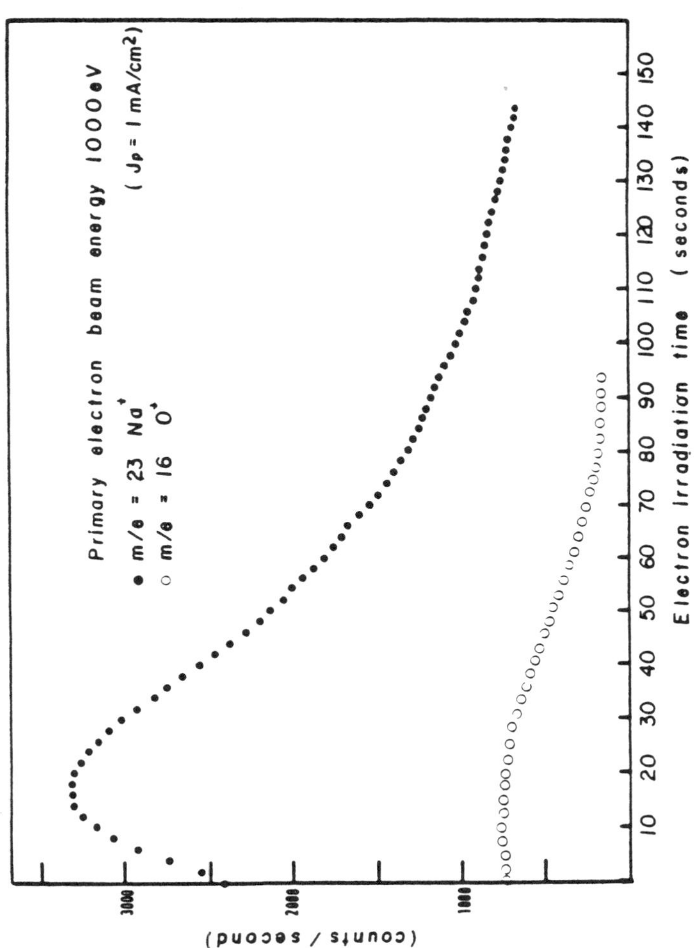


Fig. 2. A desorption rate profile of m/e = 16 and 23 signals from the soda-silica glass surface as a function of electron beam bombardment time. A 1 keV, 1 mA/cm$^2$ electron beam was used.

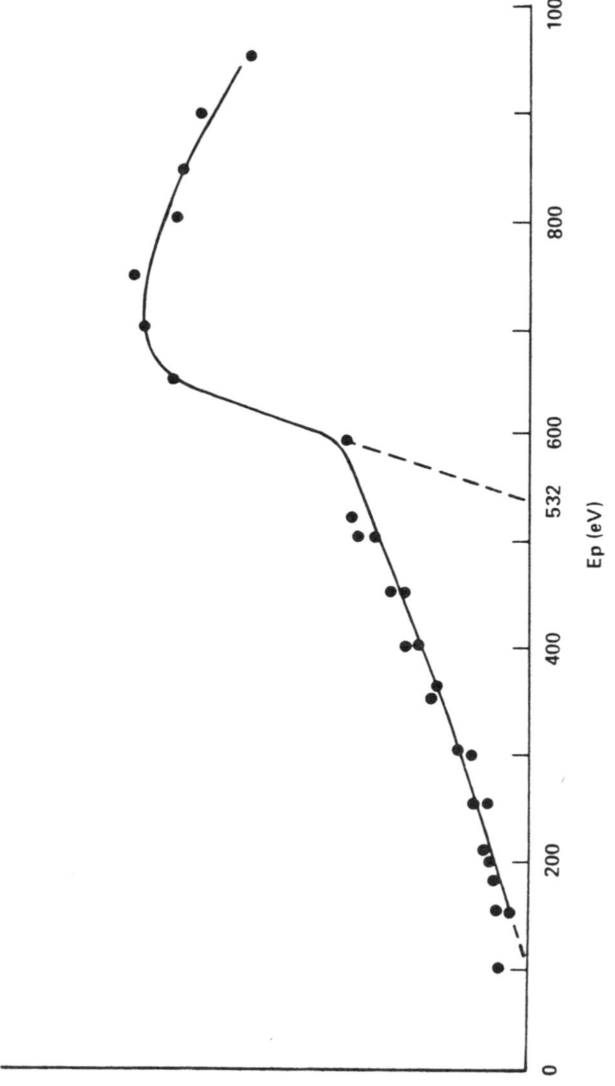


Fig. 3. Intensity of mass spectrometer signal at m/e = 23 representing $Na^+$ versus primary electron beam energy. Note the first increase at ~100 eV and second increase which projects back to ~530 eV corresponding to the critical energies for ionization of the Si 2p and O 1s levels, respectively.

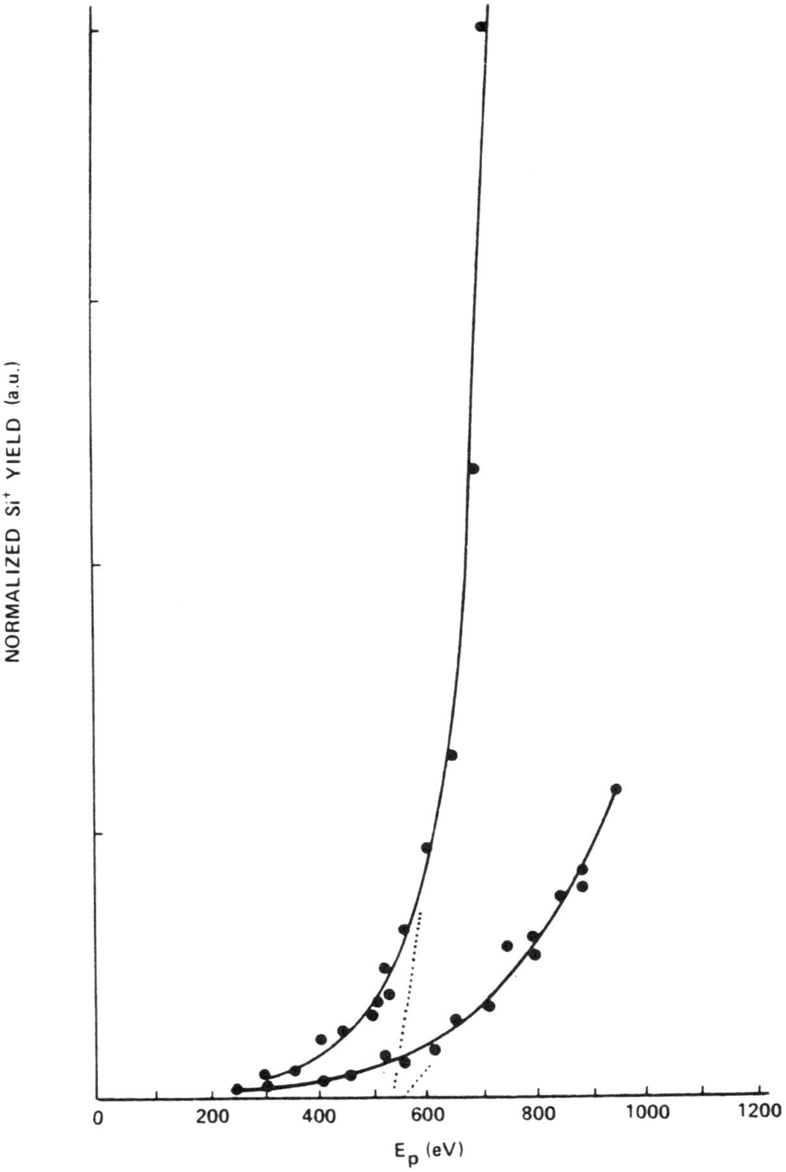

Fig. 4. Mass spectrometer signal intensity at m/e = 16 or 28 representing $O^+$ and $Si^+$, respectively, versus primary electron beam energy. The rate of desorption of both species increases dramatically when ionization can occur in the O 1s level.

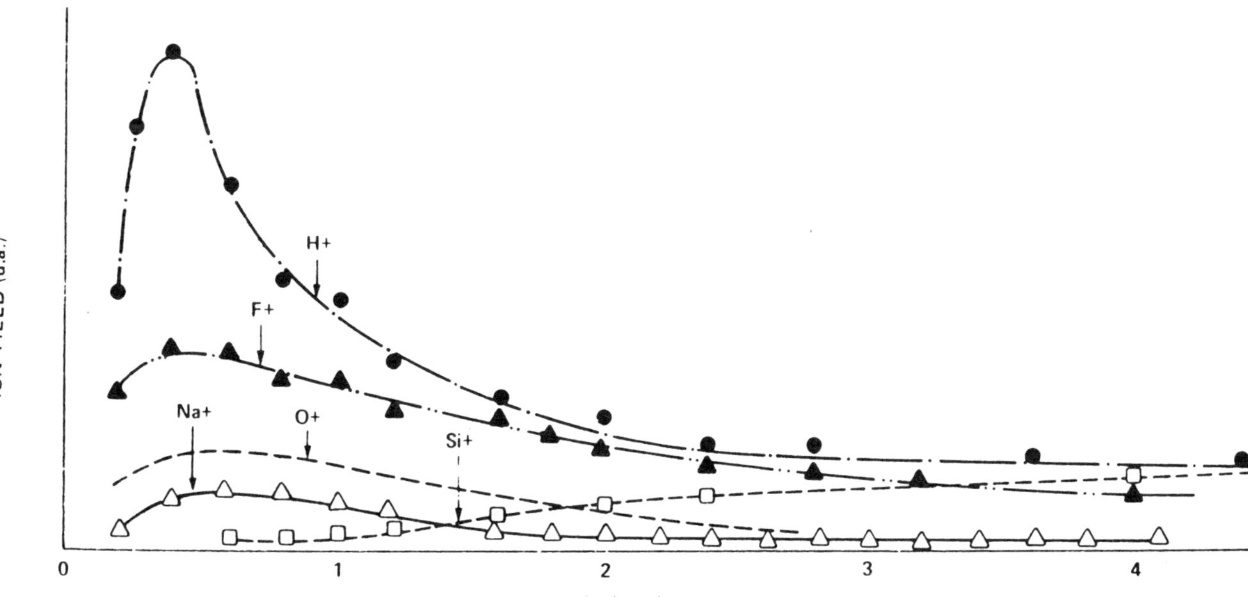


Fig. 5. Mass spectrometer signal versus electron bombardment time for m/e = 1, 16, 19, 23, and 28 representing $H^+$, $O^+$, $F^+$, $Na^+$, and $Si^+$. Note the rise and fall of the signals for $H^+$, $O^+$, $F^+$, and $Na^+$, while $Si^+$ incrases to a steady-state level at times >300s.

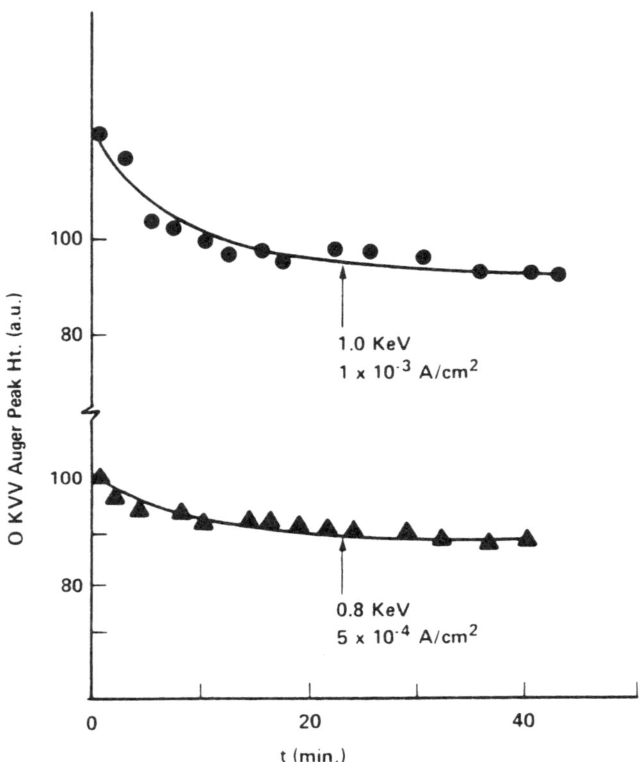


Fig. 6. Decrease in the O Auger electron peak height versus time of electron bombardment of the glass surface.

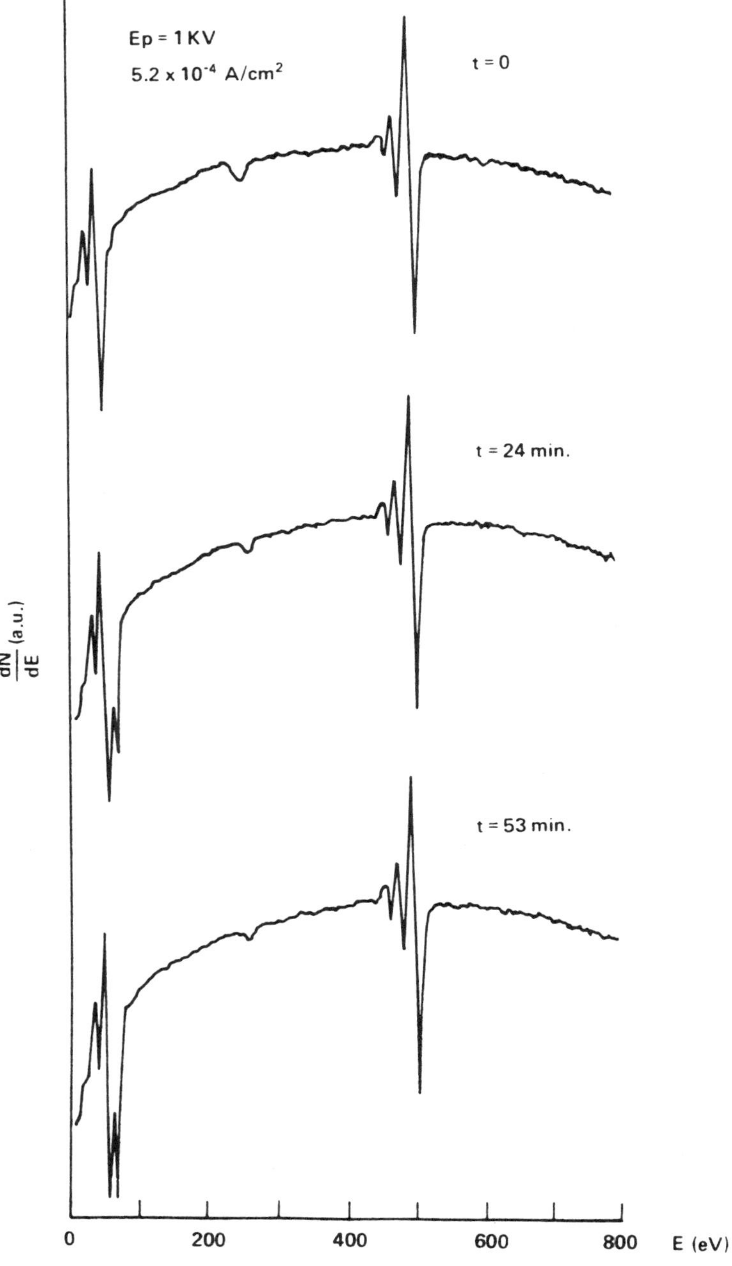


Auger Spectra from the Surface of Soda-Silica Glass.

Fig. 7. Auger spectra from the soda-silica glass with (a) no electron bombardment, and after (b) 45 min. or (c) 160 min. of electron bombardment. Note the decomposition of $SiO_2$ to elemental Si as indicated by the $L_{2,3}VV$ peak at 70 to 90 eV.

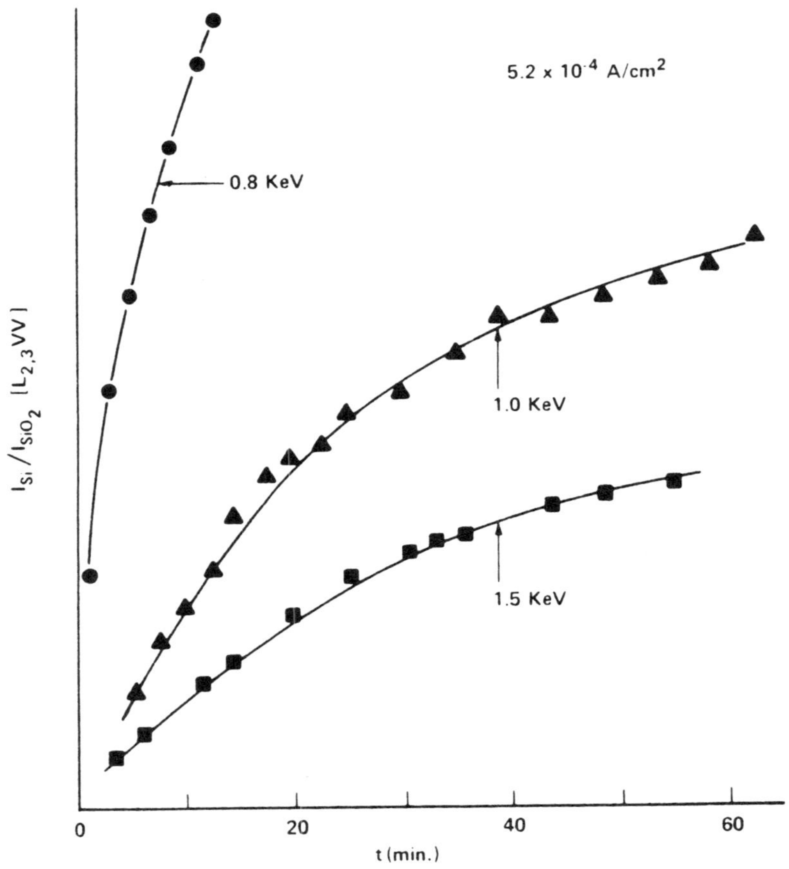


Fig. 8. Ratio of the 92 eV elemental Si to 75 eV $SiO_2$ Auger peaks versus electron bombardment time and primary beam energy for $5.2 \times 10^{-4}$ A/cm$^2$.

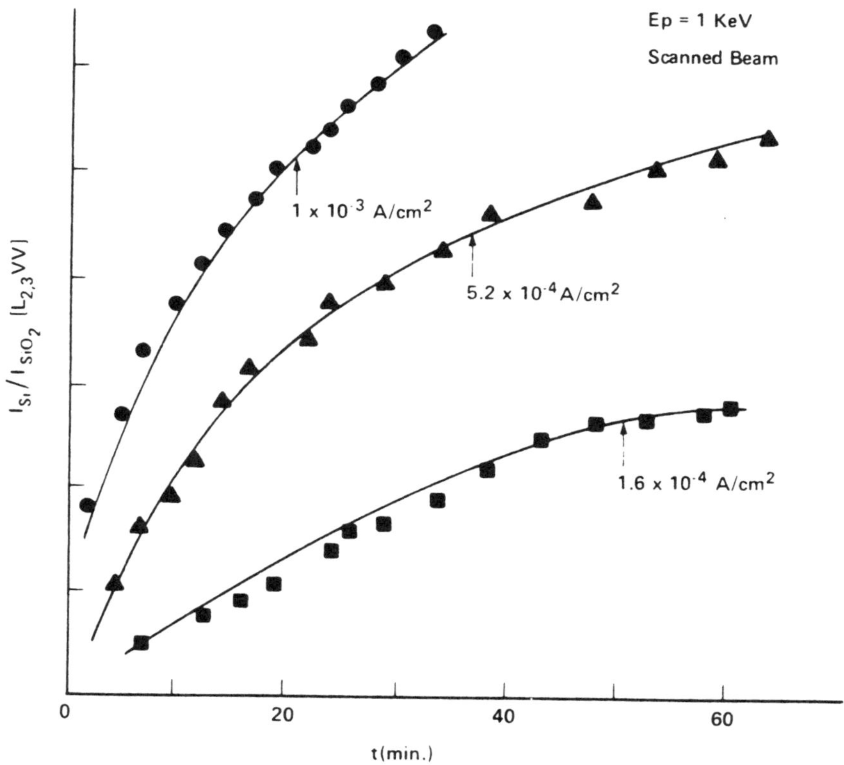


Fig. 9. Ratio of the 92 eV elemental Si to 75 eV $SiO_2$ Auger peaks versus electron bombardment time and current density for 1 keV electrons.

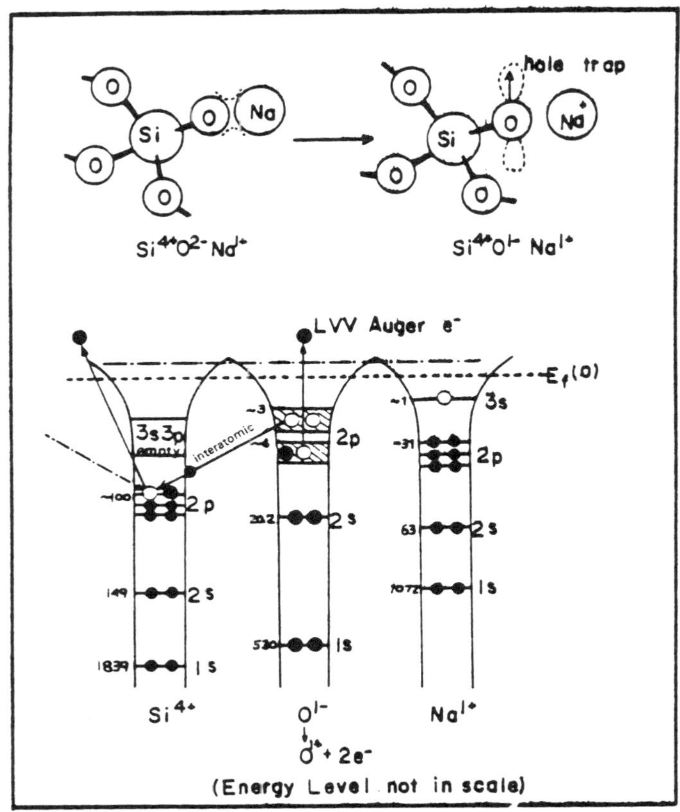


Fig. 10. Interatomic Auger de-excitation model for stimulated desorption of sodium ions. After a hole trapping in non-bridging oxygen 2p orbital of a single oxygen, formulation of a core hole in the 2p sheel of maximal valency of Si is followed by interatomic Auger decay.

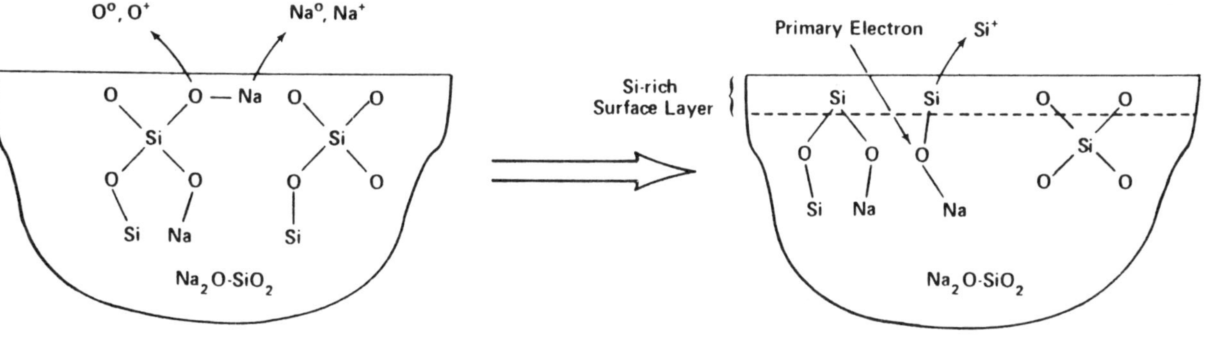


Fig. 11. Schematic representation of bonding in the glass before and after electron bombardment. Electron bombardment causes $SiO_2$ to decompose and elemental Si Auger signals are observed along with $Si^+$ desorption.

# SURFACE CHEMISTRY OF OXIDES IN WATER

G.Y. Onoda, Jr. and J.A. Casey

## I. INTRODUCTION

This paper reviews recent theoretical concepts pertaining to electrochemical reactions at oxide-water interfaces. Attention is restricted to simple systems where the aqueous solution contains ions of the solid, $H^+$ ions, $OH^-$ ions, and univalent, symmetric electrolytes. The solution is assumed to be free of multivalent ions, polyelectrolytes, surfactants, and macromolecules to limit the scope of this review.

Interfacial phenomena at oxide-water interfaces are fundamental to various phenomena in ceramic suspensions such as dispersion, coagulation, and rheology. The behavior of suspensions depends in large part on the electrical forces acting between particles, which in turn are affected directly by surface chemical reactions.

At the present time, phenomena at oxide-water interfaces are not completely understood. Theories have been modified considerably over the past 30 years to obtain closer agreement between theory and experiment. Theories have progressed to the point where the current limitations are due largely to insufficient experimental data for well-defined systems. It is hoped that experiments on the surface chemistry of monodispersed, spherical powders will overcome these limitations in the near future.

## II. EXPERIMENTAL APPROACHES FOR STUDYING SURFACE REACTIONS

The chemical reactions and the electrical characteristics of oxide-water interfaces are measured by two separate approaches. Adsorption is usually studied by solution depletion methods. Electrical characteristics are determined by electrokinetic methods such as electrophoresis, streaming potential, and electroosmosis.

### A. Adsorption Measurements

In solution depletion experiments, a concentrated solution (titrant) of a given soluble species is added in small aliquots to a suspension. The concentrations of that species existing in the suspension before and after the aliquot addition are measured. The amount of the species in solution needed to bring about this concentration change is calculated. When this amount is substracted from the amount added as titrant, what remains is the amount adsorbed on the surfaces during this change. This method is particularly suited for multivalent ions that adsorb significantly at low solution concentrations, since adsorption significantly depletes the solution.

The solution depletion method must be modified[1] to measure the adsorption and desorption of hydrogen ions. The amount of water formed in a solution when hydrogen ions react with hydroxyl ions is insignificant compared with the amount of water already present; thus it cannot be determined by solution analysis. Consequently, a complete mass balance cannot be carried out for hydrogen ions following an addition of concentrated acid. Similarly, a complete OH balance is not possible after adding concentrated base. It is possible, however, to determine changes in the difference (excess) in the amount of H ions and OH ions that occur during acid and base titrations. When $H^+$ and $OH^-$ ions combine to form water, the excess is not affected since an equal amount of both ions are involved. It is therefore this excess quantity that is monitored during pH titrations. This excess quantity in the system changes a specific amount when acid or base is added as titrant. From the change in pH, the amount of excess causing the pH change is calculated. The change in the excess at the surface is equal to the amount added by the titrant minus the amount that accounts for the pH change.

B. Electrokinetic Measurements

In electrophoresis, a particle suspended in water migrates when an electric field is applied by two immersed electrodes. The velocity of the particle is proportional to the electric field. The proportionality constant is called the electrophoretic mobility. The Smoluchowski theory in its modified form[2] relates the electrophoretic mobility to the "zeta" potential, the electric potential at a "shear" plane near the surface of the particle. The shear plane is the furthermost distance from the surface where the liquid is stagnant. Beyond the shear plane, liquid moves relative to the solid when the particle migrates.

Electrophoresis is suitable for particle sizes commonly encountered in ceramics (e.g., 0.1 to 5 µm). Streaming potential methods[3] are more suitable for coarser particles (e.g., 60-320 mesh), while electrosmosis is better for extremely fine particles such as clay.[4]

## III. NATURE OF OXIDE SURFACES

Electric charges on an oxide surface arise by the transfer of ions between the surface and the surrounding aqueous medium. This transfer is driven by chemical reactions involving active sites on the oxide surface. A net charge develops when more positive charge (via cations) are transferred than negative charge (via anions), or visa versa. The extent to which various reactions proceed depends on the concentrations of solution ions that are involved.

### A. Chemisorbed Water

Surface OH groups have been identified through infrared spectroscopy studies[5,6] as one of the most important sites for chemical reactions at oxide surfaces. These OH groups, which we shall represent by SOH, are formed by the chemisorption of water molecules on the oxide surface. The hydration mechanism, illustrated in Fig. 1, involves the dissociation of an adsorbed water molecule, where an $H^+$ ion bonds to an oxygen ion on the surface and an $OH^-$ bonds to a metal ion on the surface. It produces two SOH groups for every water molecule adsorbed. As no net transfer of charge occurs by this hydration reaction, the hydrated surface remains neutral. The surface $OH^-$ ions serve as terminal bonds for the metal ions at the surface.

### B. Effects of Thermal and Chemical History

The concentration of SOH groups on oxide surfaces depends in large part on the chemical and thermal history of the surface.[7,8] Oxide particles which have been precipitated in water or have been immersed in water for a long time usually have fully hydrated surfaces (approximately a monolayer of SOH). Upon drying of a powder, this chemisorbed water is retained. Molecular water from the surrounding air is physically adsorbed on the dry powder through hydrogen bonding to an SOH group. It can be driven off by heating the powder through the 120-200°C range. The remaining chemisorbed water remains unaffected until a temperature of around 400°C is reached. From around 400-700°C, depending on the oxide, the SOH groups react to liberate water molecules to the suroundings. Water removal continues as long as there are SOH groups on the surface that are adjacent to each other, since it requires two SOH groups to form a water molecule. When only isolated SOH groups remain, further removal of water does not occur until high enough temperatures are reached to provide sufficient surface diffusion for SOH groups. Only through migration along the surface can two SOH groups react together to form water. This occurs in temperature ranges of around 800-1100°C, depending on the oxide.

When an oxide surface is completely dehydrated, it tends to become hydrophobic. When cooled to room temperature, rehydration is slow even though it is is favored thermodynamically. Without SOH groups, water molecules lack surface sites for forming hydrogen bonds and therefore physical adsorption of water is difficult. Since the formation of SOH groups requires as a preliminary step the physical adsorption of water, rehydration is retarded. Even in liquid water, dehydrated powders often require weeks or months to rehydrate completely at the surface.

## IV. ACID/BASE SURFACE REACTIONS

The SOH groups can undergo acid and base reactions. They accept a hydrogen ion to become an $SOH_2^+$ site having a positive charge, or they release a hydrogen ion to become an $SO^-$ site having a negative charge. These reactions are written as

$$SOH + H^+(Aq) = SOH_2^+ \quad \text{(reaction 1)}$$

$$SOH = SO^- + H^+(Aq) \quad \text{(reaction 2)}$$

### A. Charge Density

The concentrations of $SOH_2^+$ and $SO^-$ species depend in large part on the pH of the aqueous phase, as can be inferred by reactions 1 and 2. The $SOH_2^+$ species increases with lower pH (higher $H^+$ activity), while the $SO^-$ species increases with higher pH. If $\Gamma_{SOH_2}$ is the number of $SOH_2^+$ species per unit area and $\Gamma_{SO}$ is the number of $SO^-$ species per unit area, the net charge density $\sigma_H$ ($\mu C/cm^2$) due to ionized SOH groups is equal to

$$\sigma_H = e(\Gamma_{SOH_2} - \Gamma_{SO}) \quad (1)$$

where e is the charge ($\mu C$) of an electron.

### B. Electrochemical Equilibrium

Reactions 1 and 2 cannot be expressed as a simple mass action law with equilibrium constants for the following reason.[9,10] The electric potential where the SOH, $SOH_2^+$, and $SO^-$ species reside is not equal to the electric potential in the aqueous solution where the $H^+$ ion resides since a net transfer of charge occurred in establishing $\sigma_H$. Because of an electric potential difference, the principles of electrochemical equilibrium must be applied, where the electrical work as well as chemical work must be considered.

For electrochemical equilibrium, necessary conditions for reactions 1 and 2 are

$$\bar{\mu}_{SOH} + \bar{\mu}_{H} = \bar{\mu}_{SOH_2} \qquad (2)$$

$$\bar{\mu}_{SOH} = \bar{\mu}_{H} + \bar{\mu}_{SO} \qquad (3)$$

where $\bar{\mu}_i$ is the electrochemical potential for species i. The electrochemical potential is divided into a chemical and electrical contribution:

$$\bar{\mu}_i = \mu_i + Z_i F\psi$$

where $\mu_i$ and $Z_i$ are the chemical potential and valence (including the sign) of species i, F is the Faraday constant, and $\psi$ is the electric potential.

A common convention is to let the electric potential of the aqueous phase be zero so that a potential $\psi$ at some location within the interface is relative to that of the aqueous phase. Following this convention, let $\psi_H$ be the average electric potential where the species SOH, $SOH_2^+$, and $SO^-$ reside. The electrochemical potentials of the four species expressed in Reactions 1 and 2 are

$$\bar{\mu}_{SOH} = \mu^o_{SOH} + RT \ln a_{SOH}$$

$$\bar{\mu}_{SOH_2} = \mu^o_{SOH_2} + RT \ln a_{SOH_2} + F\psi_H$$

$$\bar{\mu}_{SO} = \mu^o_{SO} + RT \ln a_{SO} - F\psi_H$$

$$\bar{\mu}_{H} = \mu^o_{H} + RT \ln a_{H}$$

where $\mu^\circ_i$ are the standard chemical potentials, $a_i$ are the activities, R is the gas constant, and T is temperature.

Substituting the above equations for $\bar{\mu}_i$ into equations 2 and 3 and recalling that F/R = e/k, where k is the Boltzmann constant, we obtain

$$\frac{a_{SOH_2}}{a_{SOH}\ a_H} = K_1\ \exp(-\frac{e\psi_H}{kT}) \qquad (4)$$

$$\frac{a_{SO}\ a_H}{a_{SOH}} = \frac{K_2}{\exp(-\frac{e\psi_H}{kT})} \qquad (5)$$

where $K_1$ and $K_2$ are constants equal to

$$K_1 = \exp\{\frac{\mu^o_{SOH} + \mu_H^o - \mu^o_{SOH_2}}{RT}\}$$

$$K_2 = \exp\{\frac{\mu^o_{SOH} - \mu^o_{SO} - \mu^o_H}{RT}\}$$

C. Mass-Action Relation

To utilize equations 4 and 5, the activities must be related to concentrations. Statistical mechanics arguments[9,11] lead to the conclusion that the activities of SOH, $SOH_2{}^+$, and $SO^-$ could be related, at least to a first approximation, to their mole fraction at the surface. If $\Gamma_T$ is the total density of sites given by

$$\Gamma_T = \Gamma_{SOH} + \Gamma_{SOH_2} + \Gamma_{SO} \qquad (6)$$

the activities are given by $a_{SOH} = \Gamma_{SOH}/\Gamma_T$, $a_{SOH_2} = \Gamma_{SOH_2}/\Gamma_T$, and $a_{SO} = \Gamma_{SO}/\Gamma_T$. Using these expressions, equations 4 and 5 become

$$\frac{\Gamma_{SOH_2}}{\Gamma_{SOH}} = a_H\ K_1\ \exp(\frac{-e\psi_H}{kT}) \qquad (7)$$

$$\frac{\Gamma_{SO}}{\Gamma_{SOH}} = \frac{K_2}{a_H\ \exp(\frac{-e\psi_H}{kT})} \qquad (8)$$

The charge density, given by equation 1, is determined by substituting in equations 6-8 to yield

$$\sigma_H = \frac{\sigma_T \left[a_H K_1 \exp(-e\psi_H/kT) - \dfrac{K_2}{a_H \exp(-e\psi_H/kT)}\right]}{\left[1 + K_1 a_H \exp(-e\psi_H/kT) + \dfrac{K_2}{a_H \exp(-e\psi_H/kT)}\right]} \qquad (9)$$

where $\sigma_T = e\Gamma_T$, the maximum possible surface charge density. We see from this equation that $\sigma_H$ cannot be evaluated from the pH (or $a_H$) and the constants $K_1$, $K_2$, and $\sigma_T$ alone. The reason is that there is another variable $\psi_H$, so that at least another independent equation is needed relating the same variables. This leads to the next step, the formulation of specific physical models for all of the charge sites within the interface.

## V. A SIMPLE REACTION MODEL[9-11]

The approach for incorporating acid/base surface reactions into physical models of the interface can be illustrated easiest by considering the simplest model first. The physical characteristics of the model is illustrated in Fig. 2. It consists of two distinct planes, the H layer and the D layer. The H layer is the location where the charges due to $SOH_2^+$ and $SO^-$ reside. The D layer is the distance of closest approach for counter ions, which are ions that are only attracted electrostatically to the charges in the H layer. Outside of the D layer is a diffuse layer of charge resulting from counter ions.

A. The Diffuse Layer

The diffuse layer of charge balances the net charge in the H layer. The charge density ($\mu C/cm^2$) in the diffuse layer, $\sigma_D$, and the variation of electric potential with distance from the surface is described by the well-known Gouy-Chapman theory as modified by Grahame.[12] If the aqueous solution is dominated by a symmetric electrolyte (e.g., NaCl), the charge density is given by

$$\sigma_D = -11.72\ N^{1/2}\ \sinh(\psi_d/2kT) \tag{10}$$

where N is the concentration (mole/l) of electrolyte and $\psi_D$ is the electric potential (mV) at the D layer, the distance of closest approach to the surface for the counter ions.

B. Equations of the Model

The combination of the acid/base reactions and the diffuse layer results in a model having four interfacial variables $\sigma_H$, $\sigma_D$, $\psi_H$, and $\psi_D$) and two solution variables (pH and N). The constants that provide uniqueness to each system are $K_1$, $K_2$ and $\sigma_T$ from Eq. 9.

For charge neutrality

$$\sigma_H + \sigma_D = 0 \tag{11}$$

With equations 9, 10, and 11, we have three equations relating four interfacial variables. To solve for these variables in terms of the solution variables, one more equation is needed. This is given by a capacitance equation

$$C_H = \frac{\sigma_H}{\psi_H - \psi_D} \tag{12}$$

With equations 9-12, it is now possible to solve for unique values of the four interfacial variables given a set of solution variables and system constants. This is facilitated mathematically by letting $\sigma_H$ be an independent variable and then solving for the other variables, including pH, for given N and system constants. The steps for this solution are:

(1) select a value for $\sigma_H$

(2) from equation 11 determine $\sigma_D = -\sigma_H$

(3) from equation 10 determine $\psi_D = \frac{2kT}{e} \sinh^{-1} \left(-\frac{\sigma_D}{11.72N^{1/2}}\right)$

(4) express equation 9 in root form and rearranging as

$$a_H = \left(\frac{\sigma_H + [\sigma_H^2 + 4K_1K_2(\sigma_T-\sigma_H)(\sigma_T+\sigma_H)]^{1/2}}{2K_1(\sigma_T-\sigma_H)\exp(-e\psi_H/kT)}\right) \quad (13)$$

where pH = -log $a_H$ and all variables on the right side of the equation are now known. Variations in the interface variables versus pH are obtained by selecting other value for $\sigma_H$ and repeating the above steps.

C. Features

The predicted features of this model are illustrated in Fig. 3, which gives calculated plots of $\sigma_H$ vs pH and $\psi_D$ vs pH for different ionic strengths, assuming a given set of system constants. The first distinguishable feature is that the curves for different ionic strength intersect at a common point, called the point of zero charge, and designated as $pH_{pzc}$. At this intersection point, $\sigma_D$=0 and $\psi_D$=0. It also follows from equation 11 that $\sigma_H$=0 and from equation 12 that $\psi_H=\psi_D$=0 at this point. With these conditions applied to equation 13, we can deduce that the $pH_{pzc}$ is related to the ratio of the equilibrium constants for the acid and base surface reactions by

$$pH_{pzc} = -\log(\frac{K_2}{K_1})^{1/2}$$

D. Limitations

Although the simple reaction model agrees semi-quantitatively with published results for adsorption and zeta potential isotherms, there remains several deficiencies and unexplained phenomena. One problem is the lack of quantitative agreement for the adsorption and zeta potential isotherms. The second, major problem is the time dependent effects that are often observed experimentally that are not explained by the simple model.

For the limited data that is available in which both adsorption and zeta potential information are given, quantitative agreement between theory and experiment is not found.[9,10] The experimental adsorption densities $\sigma_H$ measured by titration are considerably larger than the $\sigma_H$ calculated from zeta potential, using the Gouy-Chapman equation and assuming $\psi_D=\xi$. While it is sometimes possible to fit theory to $\sigma_H$ versus pH curves, or to $\xi$ versus pH curves, the same set of system constants cannot be used to get both types of curves to agree. If the experimental data are valid, the disagreement suggests that the net surface charge from immobilized ions is smaller than that given by $\sigma_H$. This can occur, for example, by having counter ions which partially neutralize $\sigma_H$, penetrating below the electrokinetic shear plane and becoming immobilized. This assumption is the basis for several models discussed in section VI-A.

Time dependent effects are frequently encountered in practice which can cause serious experimental difficulties. Dramatic aging effects during potentiometric titrations were first reported in detail for $Fe_2O_3$[13] and has since been reported with other oxides.[14-17] When an acid or base titrant is added, the pH of the suspension changes rapidly (within minutes) to a new value. Thereafter, the pH is found to change gradually back toward the pH existing before the titrant was added. This change can occur for weeks or months, eventually reaching some intermediate pH value. The rapid pH change followed by a drift to an equilibrium value after long times indicates that there are two surface reactions, one fast and one slow, involving hydrogen ions.

Experimentally, most adsorption isotherms by potentiometric titrations are carried out rapidly so as to suppress the effects of the slow reaction. The more rapid the titration measurements, the more reversible is the isotherm. If the titration is carried out too slowly, considerable hysteresis results, presumably because of the effects of the slow reaction.

Several hypotheses have been proposed for the fast and slow reactions. Onoda and de Bruyn,[13] studying ferric oxide, suggested that the slow step is due to diffusion of protons in or out of a diffuse layer of protons on the solid side of the interface. Berube, Blok, and de Bruyn[15,17] suggested that the slow reactions with titania and zinc oxide were due to a slow ion exchange involving a "surface phase" which binds counter ions. Trimbos and Stein[16] suggested that the slow step with zinc oxide was associated with a diffuse electron layer on the solid side of the interface.

As slow and fast reactions are observed with a wide variety of oxides, it is questionable whether any of the above proposed mechanisms could apply to all cases. It would be simpler to explain the results if a common mechanism existed.

Another time dependent effect is the slowly shifting electrokinetic $pH_{pzc}$ as certain powders are aged in water for long times (weeks and months). This effect appears to be correlated with the degree of hydration of the oxide surface.[18,19] A powder precipitated in water or aged in water for months has a certain $pH_{pzc}$ (e.g., pH 9.2 for alpha alumina). If that same powder is dehydrated at 1000°C, cooled, and reimmersed in water, the electrokinetic pzc can be different (e.g., pH 6.8 for alpha alumina) from the original value. With exposure time to water, the $pH_{pzc}$ measured periodically shifts back gradually toward the initial value.[19] This shift is attributed to the progressive rehydration of the surface.

## VI. RECENT THEORETICAL MODIFICATIONS

### A. Counter ion Penetration

The apparent problem of having too high adsorption densities for corresponding zeta potentials has lead to modified models in which the counter ions of opposite charge penetrate below the electrokinetic shear plane. In this way, the actual net charge within the shear plane is smaller than what is calculated from the hydrogen ion adsorption density. This type of model allows much greater adsorption of hydrogen ions without a corresponding large increase in $\psi_D$. Three models of this type are the gel model, the site binding model, and the specific adsorption model.

The gel model[9,20-22] assumes a surface covered with a layer of hydrolyzed material of finite thickness. Within this gel layer, acid/base reactions occur. In addition, counter ions penetrate the gel, neutralizing some of the charge due to the hydrogen ion transfer. With the outer surface of the gel serving as the electrokinetic shear plane, this model allows a considerable amount of excess or deficit $H^+$ in the gel without a large increase in the electric potential at the shear plane.

The site binding model[23-25] assumes that a counter ion of opposite charge chemically bonds to an $SOH_2^+$ or $SO^-$ site. This immobilizes the counter ion and neutralizes the surface charge. With a fraction of surface sites neutralized, the density of $SOH_2^+$ sites or $SO^-$ sites can increase considerably without a corresponding increase in $\psi_D$.

Simple specific adsorption of counter ions with a fixed number of available sites have been proposed for a planar model[26] and for a gel model.[27] The adsorption of counter ions is independent of the SOH sites and their modification.

B. Aging and Non-Equilibrium

A pH drift occurring during potentiometric titrations indicates that the suspension is not at true equilibrium. It has recently been suggested[28] that existing reaction models are incomplete because all of the conditions for equilibrium are not being satisfied. While the relations specified in these models are necessary for equilibrium, they are not sufficient if the equilibrium of some reaction was not taken into account.

What has been ignored in recent models is the equilibrium conditions for the lattice ions of the solid. The electric potential difference $\psi_o$ between the bulk solid phase and the water phase is not specified in recent models. It would be incorrect to assume that $\psi_o=\psi_H$. Rather, as discussed by de Bruyn and Berube,[14] $\psi_o$ should be given by the Nernst equation

$$\psi_o = 2.303\ (kT/e)\ (pH_o - pH) \tag{14}$$

where $pH_o$ is the pH where $\psi_o = 0$. This equation comes from the fact that $\psi_o$ is related to the activities of the lattice ions in water, which in turn are dependent on pH.

The transfer of lattice ions between the solid and water has not been considered seriously as a charging mechanism in recent models. Because of the low solubility of most oxides, it has been argued that there are not enough metal ions in solution to account for appreciable charge densities by their transfer. While this argument may be true in many cases for the metal ions, it is not valid for the oxygen ions in the solid.[28] Oxygen ions transfer to the liquid by the reaction

$$O^{-2}\ (\text{solid}) + 2\ H^+\ (\text{water}) = H_2O$$

or back to the solid by the reverse of this reaction. Whenever an oxygen ion leaves the solid, two hydrogen ions in solution are consumed. The amount of $O^{-2}$ ions transferred by this mechanism is essentially unlimited. Since a charge contribution due to lattice ions depends on the excess of metal ion charge minus the oxygen charge, a positive charge can arise by having a deficit of oxygen ions. Similarly, a negative charge arises by an excess of oxygen ions. In other words, a surface charge can develop by the transfer of the oxygen ion alone, without a significant transfer of metal ions.

When oxygen is transferred to the solid, the pH of the solution is altered in the same way as if $H^+$ ions were desorbed, since the reaction creates two $H^+$ ions in water. Conversely, the transfer of $O^{2-}$ to water has the same effect as two $H^+$ ions adsorbing to the surface. Therefore, although the $H^+$ is not actually transferred, it has the same effect on a potentiometric titration. Thus, adsorption isotherms calculated from potentiometric titrations would have a contribution from the lattice ion transfer mechanism as well as from the acid/base reaction.

If the charging mechanism by lattice ions occurs, then there are two mechanisms for creating charge, the second being the acid/base reaction. As the lattice ions lie below the H layer, the reaction involving them might be slower than the acid/base reaction, therefore accounting for the fast and slow step in potentiometric titration.

C. A Variable PZC Model[28]

This model incorporates the charging mechanism due to lattice ion transfer with the simple reaction model. Another layer, the O layer, is added to the simple reaction model, as illustrated in Fig. 4. Layer O is the layer where excess/deficit lattice ions are assumed to reside and where the electric potential $\psi_o$ is given by the Nernst equation. The charge density due to lattice ions is given by $\sigma_o$.

The added layer requires the addition of two new equations, the Nernst equation and an additional capacitance equation:

$$\psi_o = 2.303\ kT/e\ (pH_o - pH) \tag{14}$$

$$C_o = \sigma_o/(\psi_o - \psi_H) \quad (15)$$

and a modification of equations 11 and 12 to give

$$C_H = (\sigma_o + \sigma_H) / (\psi_H - \psi_D) \quad (16)$$

$$\sigma_o + \sigma_H + \sigma_D = 0 \quad (17)$$

These, along with equation 9 (the H reaction equation) and equation 10 (the Gouy-Chapman equation) gives six independent equations relating $\sigma_o$, $\sigma_H$, $\sigma_D$, $\psi_o$, $\psi_H$, and $\psi_D$ to pH and N. The method for solving these equations is presented in reference 28.

Using reasonable values for the various constants, the general features of the variable pzc model can be considered. The sum $\sigma_o + \sigma_H$, measured in potentiometric titrations, is plotted versus pH in Fig. 5a. The corresponding $\psi_D$ versus pH curves are plotted in Fig. 5b. It can be seen that both sets of curves have features quite similar to the curves for the simple reaction model (Fig. 3). However, the location of the pzc is not the same. In the reaction model, the pzc occurs at $pH_R = -2.303 \ln (K_2/K_1)^{1/2}$. In a silver iodide model[29] where surface charging is due exclusively to lattice ions, the pzc would be equal to $pH_o$. What we find in the present model is that the pzc lies between $pH_o$ and $pH_R$. The present model, because it has two charging mechanisms, is in effect a composite model between the silver iodide model and the reaction model.

Given the values of $pH_o$ and $pH_R$, the $pH_{pzc}$ depends on the ratio $C_o/\sigma_T$. The smaller this ratio, the closer is $pH_{pzc}$ to $pH_R$. The larger this ratio, the closer is $pH_{pzc}$ to $pH_o$.

Among the constants of the system, the one that can be altered experimentally is $\sigma_T$, since this is equal to the extent of hydration at the surface. When a surface is dehydrated by heating, $\sigma_T$ becomes small, and so the $pH_{pzc}$ measured for this powder would be closer to $pH_o$. As the surface gradually rehydrates upon exposure to water, $\sigma_T$ increases, causing a shift in $pH_{pzc}$ in the direction of $pH_K$, as illustrated in Fig. 6. This mechanism can therefore explain the influence of the degree of surface hydration on the experimental $pH_{pzc}$.

If the acid/base reaction is fast, while the lattice ion transfer is slow, a two step (fast and slow) mechanism would be expected in potentiometric titrations. During an acid addition, the pH is decreased while $H^+$ ions adsorb into the H layer; $\sigma_o$ does not change at first. Slowly with time, $\sigma_o$ increases, causing an abstraction of hydrogen ions from solution via reaction 1. This causes the pH to drift toward higher values. Likewise, a base addition causes an initial pH increase and rapid desorption of $H^+$ from the H layer. This is followed by a gradual formation of hydrogen ions in solution via reaction 1 in reverse, causing a drift to lower pH values.

## VII. CONCLUSIONS

The existing reaction theories and their modifications provide a sound foundation for describing the electrochemical equilibrium at oxide-water interfaces. What is needed now is more experimental work on well defined systems to test the various hypotheses of these theories. The availability of monodispersed powders of simple morphology and high purity should open the way toward more meaningful experiments. In addition, potential sources for experimental error must be scrutinized in greater detail. These include ion complexations not taken into account before and surface contaminants. The causes for time dependent effects must be clearly resolved before the interpretations based on assumptions of equilibrium can be made with confidence.

REFERENCES


1. Parks, G. A. and De Bruyn, P. L., J. Phys. Chem. 66, 967 (1962).

2. Overbeek, J. Th. G., Adv. in Colloid Science III, New York (1950_, p. 97.

3. Ball, B. and Fuerstenau, D. W., Miner. Sci. Engng., 5 (4), 267 (1973).

4. van Olphen, H., CLAY COLLOID CHEMISTRY, Interscience Publishers, NY, 1963.

5. Little, L. H., INFRARED SPECTRA OF ADSORBED SPECIES, Academic Press, New York, 1966.

6. Hair, M. L., INFRARED SPECTROSCOPY IN SURFACE CHEMISTRY, Chapt. 4, Edward Arnold, Baltimore, 1967.

7. Hair, M. L. and Hertl, W., J. Phys. Chem. 73, 2372, 4269 (1969).

8. McDonald, R. S., J. Physic. Chem., 62, 1168 (1958).

9. Levine, S. and Smith, A. L., Disc. Faraday Soc., 52, 290 (1971).

10. Wright, H. J. L. and Hunter, R. J., Austr. J. Chem., 26, 1183, 1191 (1973).

11. Healy, T. W. and White, L. R., Advances in Colloid and Interface Science, 9, 303 (1978).

12. Grahame, D. C., Chem. Revs., 41, 441 (1947).

13. Onoda, G. Y., Jr. and De Bruyn, P. L., Surface Science, 4, 48 (1966).

14. Berube, Y. G. and De Bruyn, P. L., J. Colloid Interface Science, 27, 305 (1968).

15. Blok, L. and De Bruyn, P. L., J. Colloid Interface Sci., 32, 518, 527, 533 (1970).

16. Trimbos, H. F. A. and Stein, H. N., J. Colloid Interface Sci., 77, 386, 397 (1980).

17. Berube, Y. G., Onoda, G. Y. and De Bruyn, P. L., Surface Sci., 7, 448 (1967).

18. Parks, G. A., Chem. Revs., 65, 177 (1965).

19. Robinson, M., Pask, J. A. and Fuerstenau, D. W., J. Amer. Ceram. Soc., 47, 516 (1964).

20. Lyklema, J., Electroanal. Chem., 18, 341 (1968).

21. Perram, J. W., J. C. S. Faraday II, 69, 993 (1973).

22. Perram, J. W., Hunter, R. J. and Wright, H. J. L., Aust. J. Chem., 27, 461 (1974).

23. Yates, D. E., Levine, S. and Healy, T. W., J. C. S. Faraday Trans. I, 70, 807 (1974).

24. Davis, J. A., James, R. O. and Leckie, J. O., J. Colloid Interface Sci., 63, 480 (1978).

25. Davis, J. A. and Leckie, J. O., J. Colloid Interface Sci., 74, 32 (1980).

26. Bowden, J. W., Posner, A. M. and Quirk, J. P., Aust. J. Soil Res., 15, 121 (1977).

27. Levine, S., Smith, A. L. and Brett, A. C., VI$^{th}$ Int. Congr. Surface Active Agents, Zurich, 1972 (Hanser, Munich), p. 603.

28. Onoda, G. Y. and Casey, J. A., Submitted to the J. Colloid Interface Sci.

29. Bijsterbosch, B. H. and Lyklema, J., Advances in Colloid and Interface Science, 9, 147 (1978).

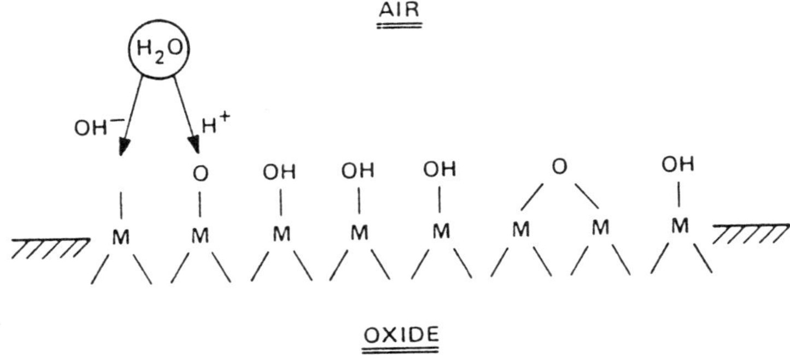


Fig. 1. Hydration state of oxide surfaces.

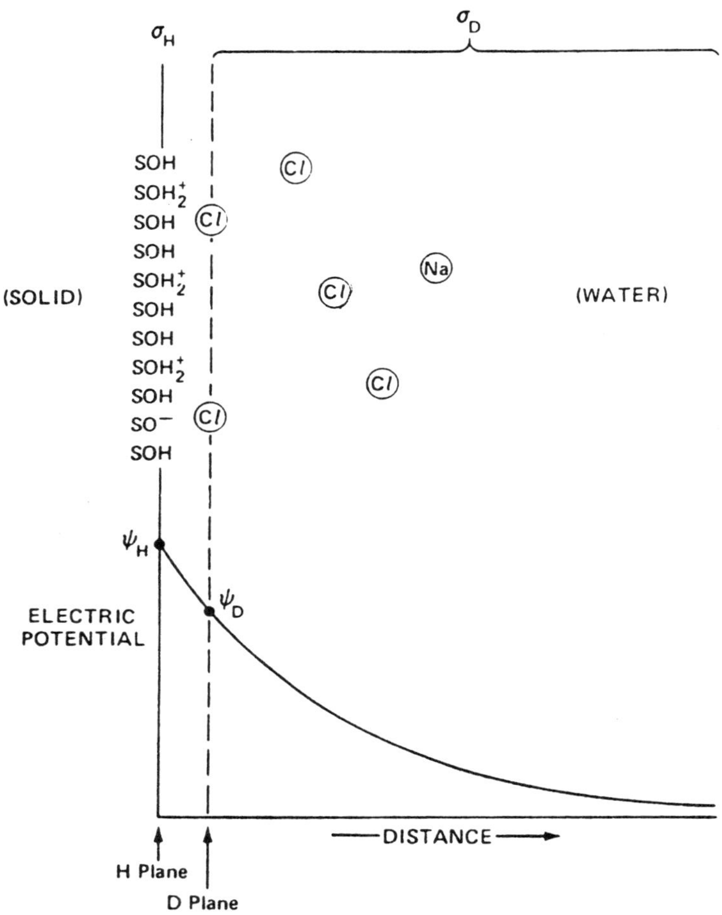


Fig. 2. Simple reaction model involving two layers (H and D).

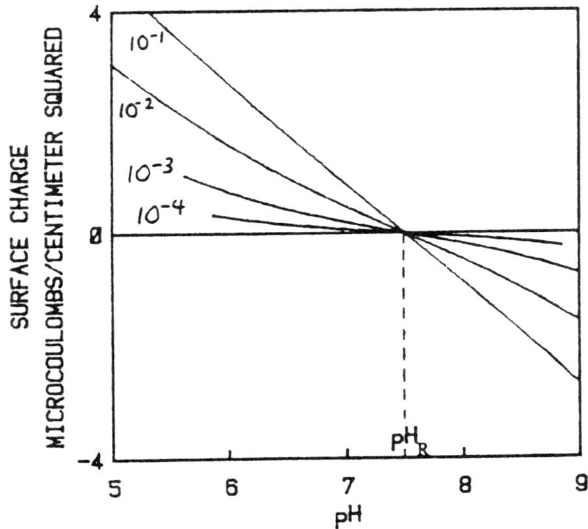


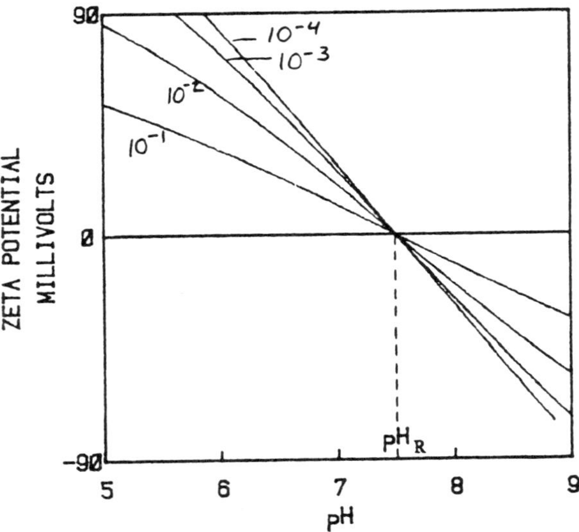


Fig. 3. Typical calculated curves for the simple reaction model.

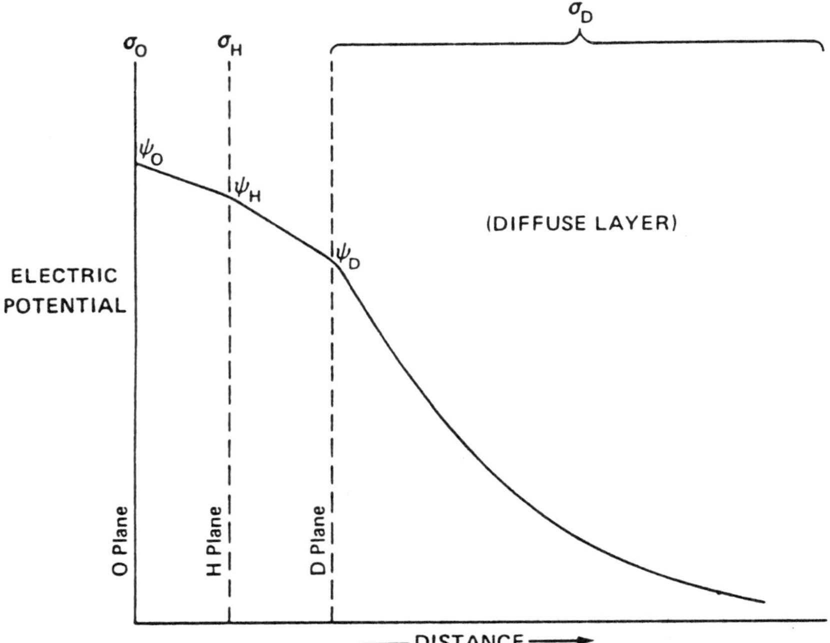


Fig. 4. Variable PZC model involving three layers (O, H, and D).

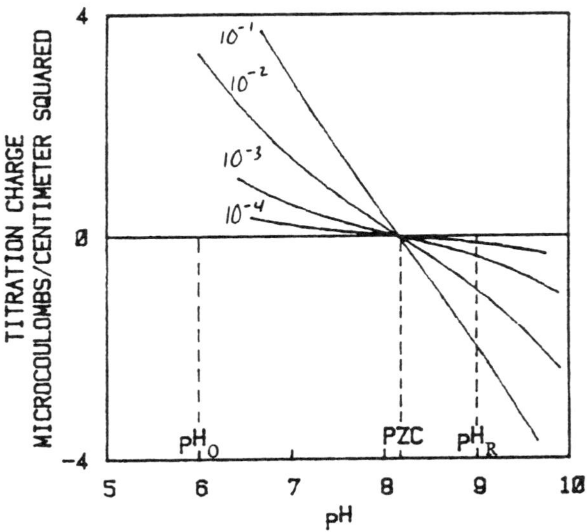


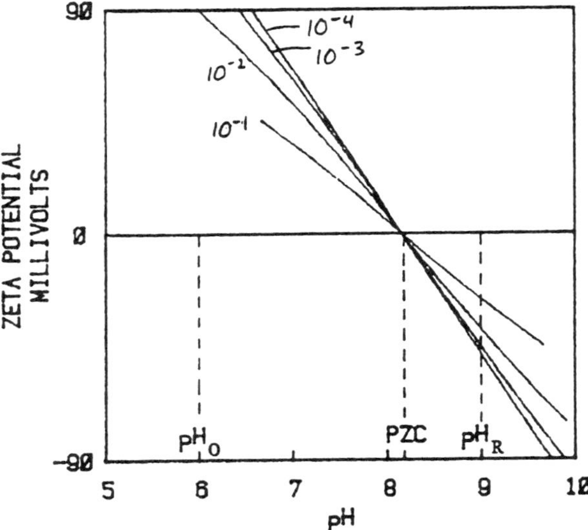


Fig. 5. Typical calculated curves for the variable PZC model.

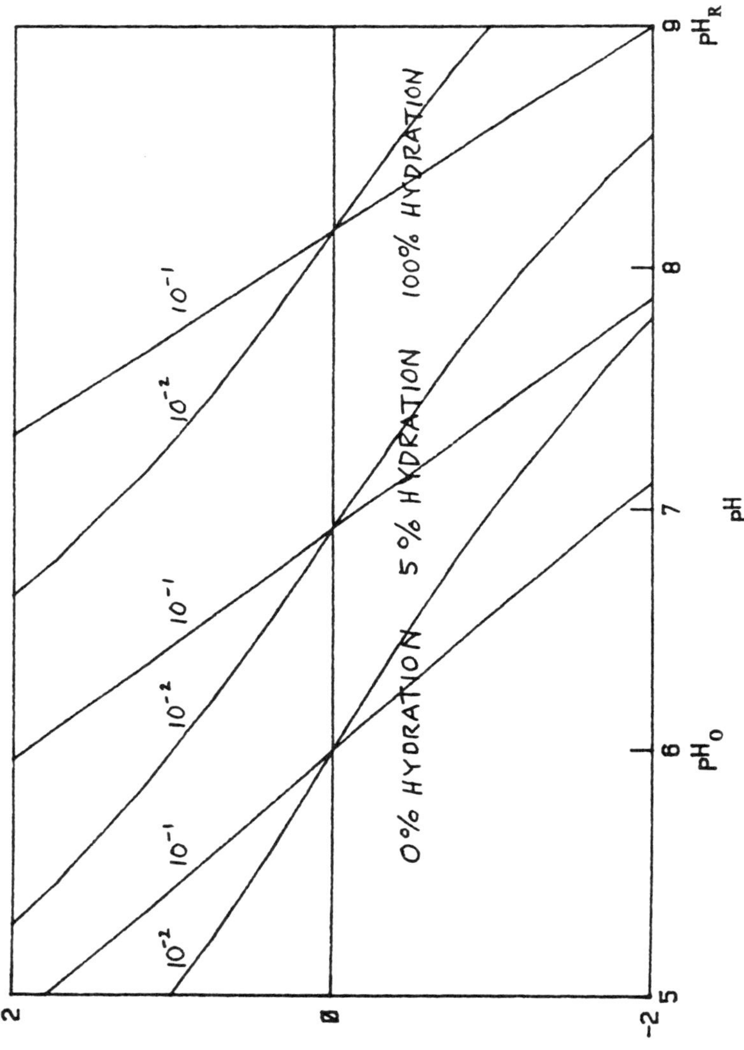


Fig. 6. Dependence of the PZC on the degree of hydration in the variable PZC model.

# SILICON NITRIDE AND SILICON CARBIDE FROM ORGANOMETALLIC AND VAPOR PRECURSORS

**Charles L. Beatty**

*Department of Materials Science and Engineering*
*University of Florida*
*Gainesville, FL 32611*

Abstract

Three different methods for producing ceramic materials are presented. One method involves the use of R.F. plasma polymerization to produce thin, conformable, crosslinked polymer coatings up to $\sim 1\mu$ thick. These polymeric materials upon pyrolysis can be converted to ceramic-like materials that ESCA analysis indicates are SiC or $Si_3N_4$ depending upon the precursor that is polymerized. Multilayer and compositionally variable thin films of these ceramics can also be produced by this technique. A second method of forming SiC involves the pyrolysis of polysilastyrene or polycarbosilane polymers. Analytical and characterization studies of these polymers indicate that they behave similar to classical organic polymers and that they will pyrolyze to SiC. These polymers can be fabricated by extrusion and other conventional polymer processing operations into complex shapes (e.g., fibers) prior to pyrolysis. The third method is a chemical vapor deposition technique for the formation of uniform, although not monodisperse, powders of SiC. It has been demonstrated that $0.2\mu$ to $10\mu$ diameter powders can be produced. The sub-micron particles may be sintered (after compacting at room temperature), into a ceramic body without application of external pressure during the sintering process.

## Introduction

My definition of ultrastructure processing is that controlled ultra small structures may be necessary for effective processing or may develop as a result of processing. The purpose of this paper is to review the work that my research group has performed during the last two years that relates to ultrastructure formation and ultrastructure processing.

This research involves three different routes to ceramics and each route yields a distinctly different geometry in the ultrastructure size ranges as indicated below:

1. Radio Frequency (R.F.) plasma polymerization of small organometallic molecules to form thin polymeric coatings that are pyrolyzable to ceramic-like coatings.
2. The characterization and pyrolysis of polymeric precursors to silicon carbide in either the coating or fiber form.
3. Use of a chemical vapor deposition process to produce sub-micron to $\sim 10\mu$ uniform SiC particles.

## R.F. Plasma Process

The R.F. plasma technique has been widely utilized to produce thin, crosslinked organic polymer films that are generally uniform in thickness even on rough surfaces (1). The uniformity of thickness is most often attributed to vapor phase polymerization that is thought to occur in R.F. plasma polymerization. In this particular system, the inductively coupled reactor (Figure 1) is operated at sufficiently low power levels ($\leq 100$ watts) that the substrate and reactor remain nominally at room temperature (2). However, the plasma produces a mixture of electrons, ions and ultraviolet radiation that is extremely reactive. If fact, these low power R.F. plasmas are sufficiently reactive that species may be polymerized that cannot normally be polymerized by conventional organic chemistry. Naturally this means that expensive, tailored monomers are not necessary. However, R.F. plasma polymerization studies on somewhat exotic chemical structures (Figure 2) as starting materials will be reported here.

The variables that influence the R.F. plasma polymerization process include: the operating pressure of the reactant and/or any inert vapor phase, the chemical composition of the substrate, the location of the substrate in the reactor, the R.F. power level and the time of polymerization. The time of polymerization and the R.F. power level are particularly important variables as there are always two basic processes occuring during R.F. plasma polymerization. One, the fragmentation of the reactant molecule by the R.F. plasma followed by recombination (often accompanied by a rearrangement) into a crosslinked polymeric network. Second, the plasma can also interact with the polymer film that has already been deposited. This concurrent modification process can for selected polymer systems result in drastic changes in chemical structure as well as a non-linear dependence of polymerization rate on power level. In fact, even a maximum in deposition rate can be observed as power level is increased for some systems (3). However, as illustrated in Figure 3, all the organometallic starting materials yielded linear increases in weight % R.F. plasma polymerized polymer versus polymerization time as R.F. power level is increased. This suggests that the modification process is not very significant for the starting materials. Representative data for the dimer and trimer of silizane is presented in Table 1.

Although initially uniform conformable coatings appear to occur on both stainless-steel foil and sodium borosilicate glass slide surfaces, R.F. plasma polymerization of thick films (Figure 4) indicate that a non-uniform, rough surface occurs. The origin of this roughening effect is unknown but it appears to be related to instabilities in deposition that may be related to the surface energy of the substrate. An obvious route for eliminating this problem (and the one we chose) is, of course, to select a substrate that results in smooth, uniform coatings.

An extremely attractive route offered by the R.F. plasma polymerization process is the possibility that gradients of chemical composition can be achieved by varying the relative concentrations of a second reactant during the deposition process. The contact angle data in Figure 5 clearly illustrate that the surface energy of the R.F. plasma deposited film can be varied simply by adding gas phases such as nitrogen or ammonia to the R.F. plasma reactor (5). The tabulated data (Table 2) illustrate the changes in critical surface tension as well as the water limiting value of the contact angle. The variation of these values with thickness of film formed indicates the effect of either surface roughening or variation of chemical structure of the surface layer as the R.F. plasma polymerization proceeded.

The R.F. plasma polymerized organometallic films are highly cross-linked and are totally insoluble in solvents. In addition, the sorption of solvent does not result in debonding or delamination of the R.F. plasma polymerized film from the substrate. It is our belief, although not proven, that the reactive species in the plasma (i.e., electrons, ions, and UV radiation) clean and activate the substrate surface sufficiently to achieve some chemical bonding between the R.F. polymerized polymer and the substrate. Although this mechanism is speculative it certainly is consistent with the excellent adhesion possessed by these R.F. plasma polymerized coatings.

The high crosslink density and excellent adhesion of these R.F. plasma polymerized films suggest that pyrolysis may result in coherent, well-adhered ceramic films on the substrate. That, in fact, does occur, which permits the characterization of the coating as a function of the pyrolysis time, temperature and environment. As shown in Figure 6, the Fourier Transform Infrared (FTIR) spectra taken after pyrolysis at suceedingly higher temperatures demonstrate the conversion from the polymeric to the ceramic state. This conversion is better illustrated by taking the ratio of the absorbance of NH at 1180 $cm^{-1}$ to the absorbance of SiN at 924 $cm^{-1}$. The FTIR data clearly demonstrate that the NH structure of silizane disappears and is replaced by the SiN linkages that are present in silicon nitride (5).

The chemical composition and bonding of the R.F. plasma polymerized films, including the effects of pyrolysis, have been studied by ESCA (or XPS) (6). The XPS spectra of commercially available SiC (Figure 7a) and $Si_3N_4$ (Figure 7b) was taken as a means of calibration and for comparison with the pyrolyzed end-product. Note that oxygen and carbon are both present at levels that indicate these commercial materials are not pure. The two silicon peaks for SiC may be due to SiC and elemental Si but it is more probable they are due to SiC and $SiO_2$. The pyrolysis process can be conveniently followed by XPS as illustrated in Figure 8. It is worth noting that we have nearly enough control of the R.F.

plasma polymerization conditions that ceramic-like materials can be produced directly in the R.F. plasma without pyrolysis.

A major advantage of XPS is that the number of a particular type of bound atom can be determined quantitively. Thus, in principle, we can count the number of atoms at the binding energy of the bonds characteristic of SiC and $Si_3N_4$ to determine the purity as well as the nature of the ceramic that is formed. Unfortunately, carbon and oxygen are contaminants common to all the specimens so this analysis is not quite so straightforward. However, the binding energy data (Table 3) can be "calibrated" via the carbon binding energy to draw some conclusions concerning the nature of the pyrolyzed R.F. plasma coatings. The films formed initially from tetramethylsilane, TMS, result in a structure that appears to be intermediate between SiC and $SiO_2$ whereas the films formed from hexamethyldisilizane, HMDS, appear to approximate silicon nitride rather well.

In addition to characterizing the type of ceramic formed, XPS has also proved valuable in improving our RF plasma polymerization technique. Figure 9 illustrates the changes in the oxygen binding energy as a function of argon purging (to flush oxygen out of the reactor) and purging followed by scavenging residual oxygen with hydrogen in the argon carrier gas.

An interesting and hopefully practical application of ceramic coatings relates to some of their unique properties. Clearly, the R.F. plasma polymerized polymer films exhibit excellent adhesion. What about the ceramic coatings formed from them? Such a ceramic coating on 0.02 mil stainless steel shim stock was quenched from room temperature into liquid nitrogen repeatedly without delamination, although delamination was expected due to the differences in coefficient of expansion of the steel and of the ceramic coating. Subsequently, the ceramic coated steel was folded back on itself and the quenching in liquid nitrogen was repeated again without delamination. Next the sample was folded back on itself in the reverse direction and again quenched in liquid nitrogen from room temperature without any apparent delamination. Although no gross delamination was observable by the naked eye, microcracking of the ceramic coating might have occured. This was checked by placing the straightened-out test piece in a bath of saturated salt water with an uncoated piece of the same substrate placed in a separate saturated salt bath. After two months there was no indication of corrosion on the coated steel substrate indicating that microcracking did not occur. The uncoated steel was totally covered with rust in less than one week. Patent protection is being sought on the R.F. plasma process for producing ceramic coatings as well as for producing anti-corrosion barriers that allow some deformation and are thermal-shock resistant.

Another important ultrastructure characteristic of the R.F. plasma process is that multilayer laminates of different ceramics can be fabricated with controlled uniform thickness in the range of 500 Å to 10,000 Å for each layer. The properties of such multilayers have not been investigated but analogous polymeric composites possess quite different and valuable properties compared to bulk combinations of the same materials.

Probably the most far-ranging application of the R.F. plasma ceramic coating route in ultrastructure processing lies in the area of providing enhanced interfacial strength for ceramic-ceramic and glass-ceramic composites. These concepts have been applied in polymeric composites as illustrated in Figures 10 and 11. The tensile properties (i.e., strength and modulus) can either be increased or decreased by increasing or decreasing the quality of bonding between phases. This difference is large (e.g., an order of magnitude change in tensile strength) and is achieved by a very thin (∿2000 Å) interfacial coating deposited by the R.F. plasma process (7). Dynamic mechanical spectroscopic results of these composites are summarized on the transition map in Figure 11. It is worth noting that the glass transition temperature of the composite with strong interfacial adhesion is substantially higher than the unmodified filler surface composite whereas the coating that produces a weak adhesive layer has the lowest composite glass transition temperature. There is no guarantee that trends of a similar magnitude will occur in ceramic-ceramic composites but the direction of the trend in polymer-ceramic composites is unmistakable.

<u>Characterization of Polymeric Precursors of SiC</u>

As discussed previously, (8) poly(silastyrene) and poly(carbosilane) are two polymeric routes for producing silicon carbide. Poly(silastyrene) of variable dimethyl and methyl-phenyl composition was supplied by Professor West, University of Wisconsin and a sample of polycarbosilane was provided by Dr. C. L. Shilling of Union Carbide Corporation. Previously personal experience with poly(dimethylsiloxane) systems suggested that the competition between pyrolysis and degradation reactions via chain scission might be influenced by the metal impurity level. Particle (proton) Induced X-ray Emission (PIXE) measurements indicate that the poly(silastyrene) sample contains iron as its major metal contaminant at a level of approximately 50 ppm with a few other metals present at lower concentrations as indicated in Figure 12a. The poly(carbosilane) was contaminated to about the 2000 ppm level of iron (Figure 12b) with relatively larger amounts of other metals compared to the poly(silastyrene). Another sample of poly(carbosilane) has been supplied that visually appears to be less contaminated but PIXE analysis has not been completed. Study of the poly(carbosilane) materials of different metal contamination levels may be helpful in understanding both the crosslinking and pyrolysis reactions.

The poly(silastyrene) family of polymers has been studied as a function of chemical composition (Table 4) ranging from the phenylmethyl homopolymer to a supposedly random copolymer that is composed of over three fourths (i.e., 0.78) silicon atoms in the chain having dimethyl substitution. An obvious discrepancy in our data compared to the data of West (8) is in molecular weight. The almost order of magnitude reduction in molecular weight of our samples may be due, in part, to chain scission due to degradation via UV radiation. Samples #1 and #2 contained microgels of a size that was not readily removable by filtering hence accurate molecular weights could not be achieved. All samples were reprecipitated at least three times prior to the intrinsic viscosity measurements.

Dielectric relaxation spectroscopy was utilized to determine if poly(silastyrene) exhibited normal organic polymer relaxation behavior. The dielectric tan $\delta$ versus temperature data (Figure 13) indicate that the glass transition relaxation process shifts to higher temperatures as the measurement frequency is increased. However, it also appears that as the frequency of measurment is increased the single relaxation process observed at low frequency (i.e., 10kHz) begins to split into two relaxation processes which is only barely discernable in the 100 kHz data. The dielectric relaxation data is summarized for samples 1 through 4 on the transition map (Figure 14). Sample 1 (n = 0.78) and sample 4 (n = 0.1) both appear to exhibit the curvilinear relationship that is typical of Williams-Landel-Ferry (WLF) behavior (i.e., glass transition behavior). It is tempting to assume that the glass transition curve for the transition shifts systematically as the composition of the copolymer is varied with a concomittant smooth change in activation energy at each measurement frequency (i.e., slope of the curve). However, due to the uncertainty in the molecular weights of these systems, this type of analysis is questionable at this time. The glass transition temperature as determined by Differential Scanning Calorimetry (DSC) agrees well with the dielectric relaxation data for samples #1 and #4. The DSC Tg values for samples #2 and #3 appeared to occur at substantially higher temperatures than expected.

Poly(silastyrene) can be extruded into fibers as illustrated in Figure 15a which upon pyrolysis coverts into a silicon carbide type structure as determined by FTIR (Figure 15b). The weight loss experienced upon pyrolysis of the poly(silastyrene) samples (Table #5) appears to be significantly higher than expected. This may be due to incomplete crosslinking prior to pyrolysis. The Union Carbide poly(carbosilane) appears easier to crosslink with a lower loss in weight upon pyrolysis. It is worth noting for comparison purposes that the weight loss upon pyrolysis of the R.F. plasma polymerized films is normally in the 10 to 30 wt % loss region.

CVD Process for Producing Small, Uniform Silicon Carbide Powders

The basic system utilized for producing SiC powders from silanes is illustrated in the schematic block diagram in Figure 16. The diameter of particles produced in this simple apparatus can be varied from supra-micron (up to 10μ) to sub-micron (∿0.2μ) by control of process parameters. Scanning electron micrographs of particles in both size ranges (Figure 17a and 17b respectively) indicate that uniform (but not monodisperse) particle size is achievable. It should be noted that the sub-micron powders exhibit a greater tendency to agglomerate than do the supra-micron powders. Optical micrographs of large diameter powders indicate that the particles have a green color which has been inferred to mean a lack of free carbon or graphite and reasonably high purity.

The sub-micron powder (Figure 17b) was compacted at room temperature, removed from the press and sintered at 1500°C as a free-standing body at ambient pressure. It is clear from Figure 18 that sintering has occurred as the grain diameter is approximately four times the starting powder diameter. The possibility of achieving a process patent for this route to uniform SiC powder is being pursued.

Summary

It has hopefully been demonstrated that there are varied routes to achieve ultrastructure as well as ultrastructure processing. Each of the three routes that have been discussed are still basically in their infancy with each route exhibiting its own unique properties or benefits. The two polymeric routes differ considerably in the polymer synthesis and crosslinking steps but are rather similar in the pyrolysis step.

Specifically, it has been demonstrated that:

(1) R.F. plasma polymerization followed by pyrolysis results in well-bonded SiC or $Si_3N_4$ ceramic-like coatings on substrates.

(2) Poly(silastyrene) appears to exhibit typical variations in the glass transition temperature as a function of frequency and composition analogous to common organic polymers.

(3) Fibers of poly(silastyrene) can be formed and pyrolyzed to SiC.

(4) A simple CVD process can be used to produce uniform SiC particles ranging in size from 0.2μ to 10μ dependent upon process variables.

(5) Sub-micron SiC powder can be compacted at room temperature and subsequently sintered at atmospheric pressure to form a sintered ceramic body.

## Acknowledgments

The author wishes to acknowledge the financial support of this work by the Air Force Office of Scientific Research; Contract No. F49620-80-C-0047 and the Microstructure Properties of Materials Center of Excellence, University of Florida. In addition contributions by Dr. S. K. Varshney, K. Shih, Dr. K. Fujioka, Dr. P. E. Bierstedt (DuPont Corp), Dr. H. Van Rinsvelt, Dr. C. D. Batich and Dr. R. West in this research program is gratefully acknowledged.

## References

1. H. Yasuda, J. Polymer Sci., Macromol. Reviews, 16, 199 (1981).

2. S. K. Varshney and C. L. Beatty, Ceramic and Engr. Sci. Proceedings, 2, 443 (1981).

3. K. Shah and C. L. Beatty, submitted to Plasma Chemistry and Plasma Processing.

4. S. K. Varshney and C. L. Beatty, Organic Coatings and Applied Polymer Science Proceedings, 46, 127 (1982).

5. S. K. Varshney and C. L. Beatty, Organic Coatings and Applied Polymer Science Proceeding, 47, 151 (1982).

6. C. D. Batich, C. L. Beatty, P. E. Bierstedt and S. K. Varshney, Organic Coatings and Applied Polymer Sci. Proceedings, 46, 134 (1982); also submitted for publication to Ind. and Engr. Chemistry.

7. K. Shah and C. L. Beatty, Organic Coatings and Applied Polymer Sci. Proceedings, 46, 232 (1982); also submitted to J. Appl. Polymer Sci.

8. R. West, in proceedings of Ultrastructure Processing of Ceramics, Glasses and Composites, L. L. Hench and D. R. Ulrich, eds., J. Wiley and Sons, Inc. (1983).

| Polymer | Time of Polymerization (mins) | Amount of Polymer Deposited ($\mu g\ cm^{-2}$) | RF Power (watts) | Thickness of Deposited Layer ($\mu m$) | Rate of Polymerization ($\mu g\ cm^{-2}\ min^{-1}$) |
|---|---|---|---|---|---|
| PHMCTS | 60 | 43.5 | 100 | 0.472 | 0.46 |
| PHMCTS | 60 | 36.0 | 75 | 0.391 | 0.32 |
| PHMCTS | 60 | 34.5 | 60 | 0.375 | 0.31 |
| PHMCTS | 60 | 30.5 | 40 | 0.338 | 0.27 |
| PHMDS | 60 | 72.35 | 100 | 0.778 | 0.95 |
| PHMDS | 60 | 43.4 | 60 | 0.467 | 0.63 |

Table 1. Plasma polymerization of hexamethylcyclotrisilazane and hexamethyldisilazane over stainless steel foil at 200 micron Hg.

| System | Thickness- m | $\gamma_c$dyn/cm | $^{\theta}H_2O$ Limiting Valve |
|---|---|---|---|
| PHMDS | 0.08 | 24.0 ± 1.3 | 94 ± 1 |
| PHMDS | 0.177 | 27.0 ± 1.0 | 93 ± 1 |
| PHMDS | 0.233 | 28.0 ± 1.6 | 95 ± 2 |
| PHMDS | 0.532 | 27.5 ± 1.1 | 95 ± 2 |
| PHMDS-$N_2$ | 0.930 | 22. ± .05 | 92 ± 2 |
| PHMDS-$NH_3$ | 0.072 | 17 ± 1.0 | 90 ± 1 |
| PHMDS-$NH_3$ | 0.087 | 20.5 ± 1.0 | 86 ± 2 |
| PHMDS-$NH_3$ | 0.550 | 18 ± 0.6 | 84 ± 2 |
| PHMDS-$NH_3$ | 1.96 | 18 ± 1.3 | 87 ± 1 |

Table 2. Zisman critical surface tension ($\gamma_c$) and water contact angle on PHMDS film on stainless-steel foil.

| Material | Si(2p) | C(s) | N(s) | O(1s) |
|---|---|---|---|---|
| Si (a) | 98.0 | (285.0) | --- | --- |
| SiC (b) | 99.0 (+103.1) | 282 (+285) | --- | 532.8 |
| $SiO_2$ (c) | 103.1 | (285.0) | --- | 533.3 |
| $Si_3N_4$ (b) | 102.0 | (285.0) | 397.8 | 532.9 |
| PP films | | | | |
| "SiC" (d) | 101.4 | (c.285.0) | c.400 | 532.2 |
| "$Si_3N_4$ (e) | 102.3 | (285.0) | 398.0 | 532.5 |

(a) F. Grunthaener (1981)

(b) P. Bierstedt (unpublished)

(c) P E reference

(d) film #13 (TMS)

(e) film #7 (HMDS)

Table 3. Binding energies ($E_B$) in eV relative to "adventitious" carbon set at 285.0 eV.

| SAMPLE | N/M | M. W. | C. S. T. |
|---|---|---|---|
| #1 | 0.78 | --- | 28.5 |
| #2 | 0.4 | --- | 27.1 |
| #3 | 0.2 | 32,700 | 31.7 |
| #4 | 0.1 | 63,200 | 33.3 |
| #5 | N=0 | 6,400 | 34.6 |

Table 4. Characteristics of the poly(silastyrenes) that are reported in this work. N is number of dimethyl groups, and M is the number of phenylmethyl groups pendat to the silicon chain; M.W. is the average molecular weight as measured intrinsic viscosity and CST is the critical surface tension in dyne/cm.

| sample | weight loss percent (%) |
|---|---|
| soluble solid polycarbosilane (7986-51) | 67 |
| polysilastyrene (n/m = 0.78) | 82 |
| polysilastyrene (n/m = 0.20) | 85 |

Table 5. Weight loss experienced upon pyrolysis of polycarbosilane and polysilastyrene.

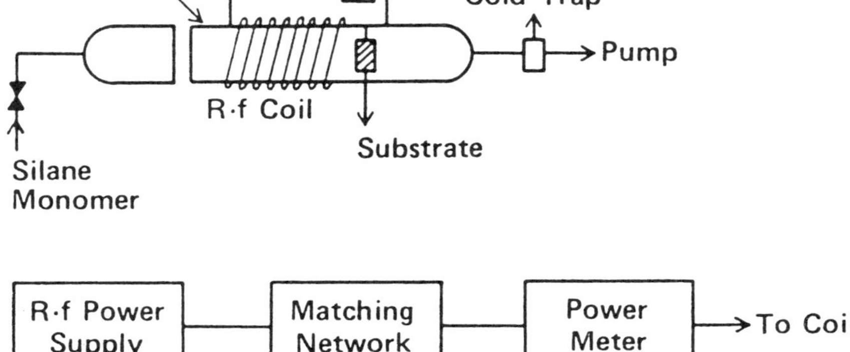


Figure 1. Schematic of the R.F. Plasma Polymerization Reactor System.

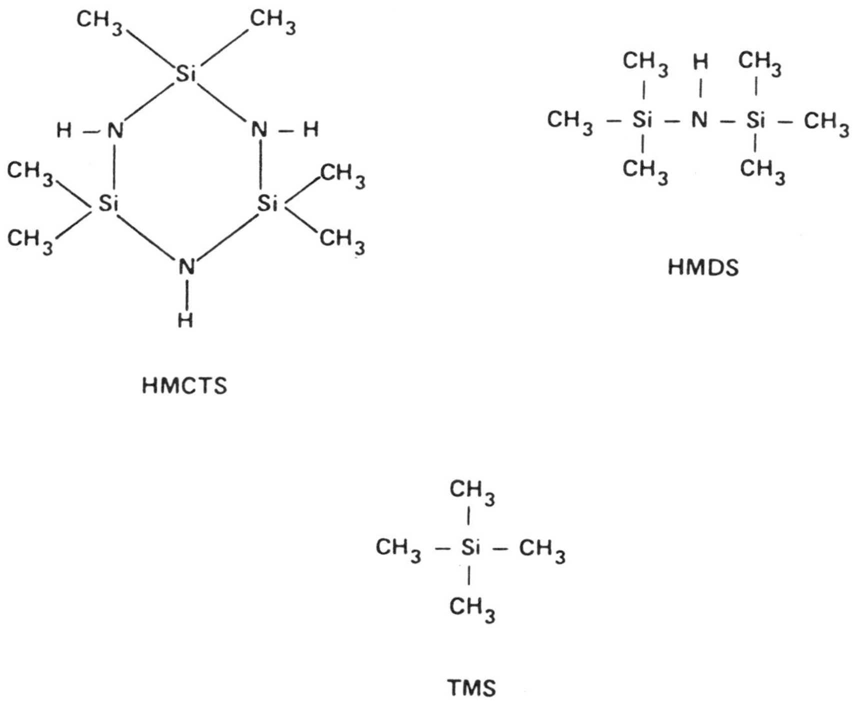

HMCTS

HMDS

TMS

Figure 2. Chemical structure of the reactants for R.F. Plasma Polymerization.

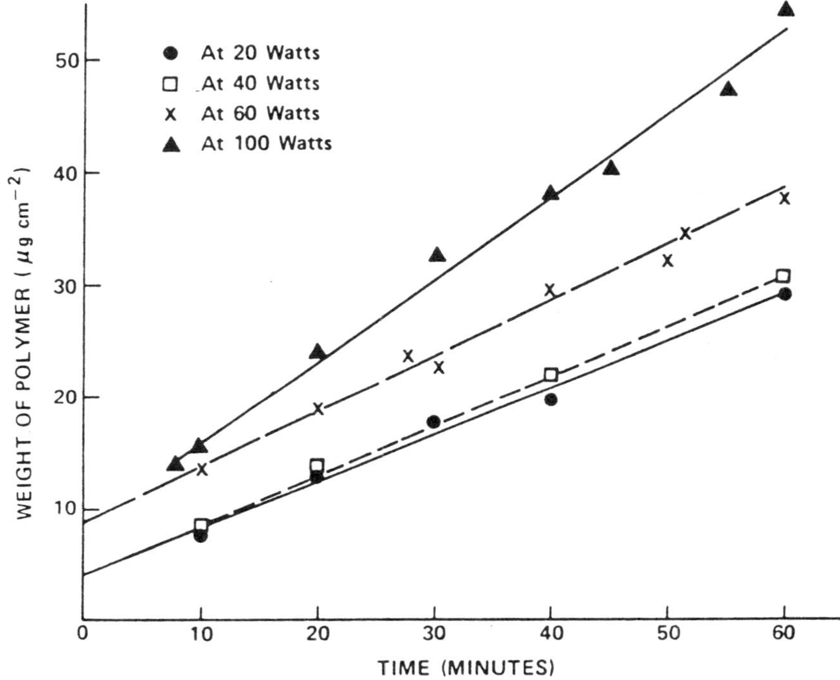


Figure 3. Kinetic of Plasma Polymerization of tetramethylsilane over stainless steel plate at 200 micron Hg

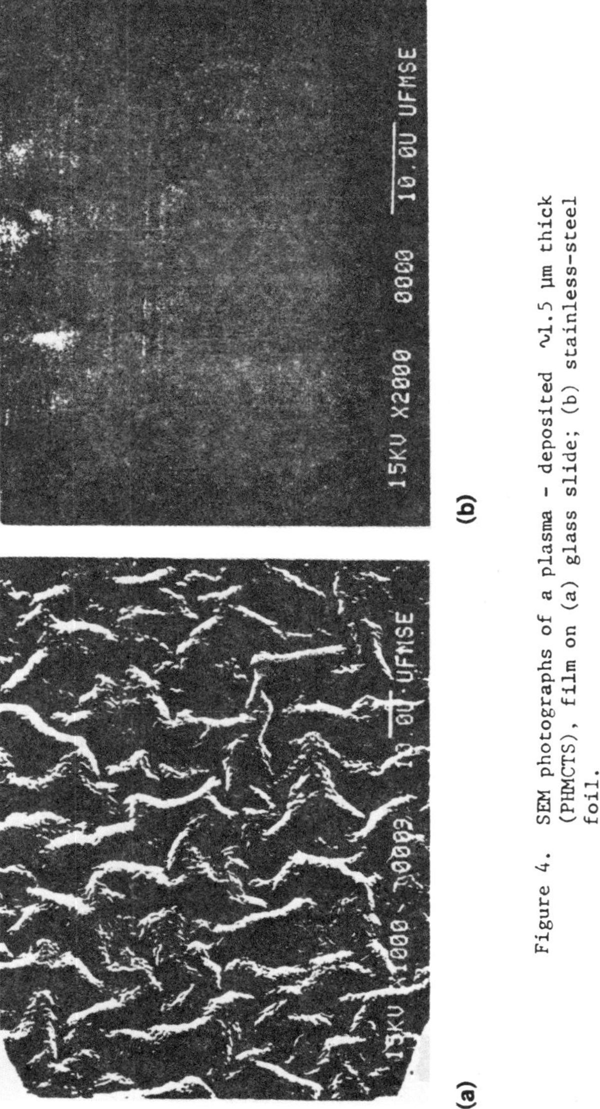


Figure 4. SEM photographs of a plasma - deposited ∿1.5 μm thick (PHMCTS), film on (a) glass slide; (b) stainless-steel foil.

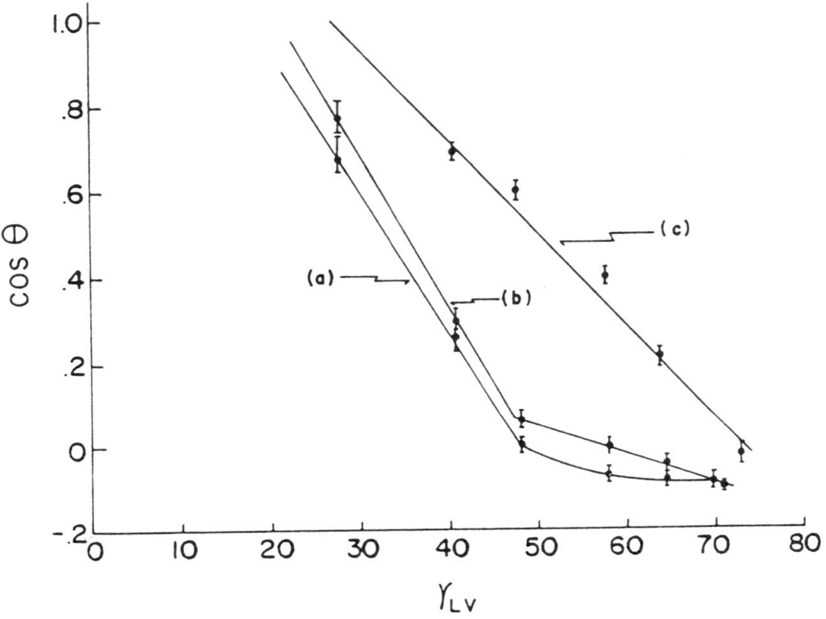


Figure 5. Zisman plot for PP - HMDS deposited on stainless-steel foil in presence of (a) ammonia (0.55 μm thick), (b) nitrogen (0.53 μm thick) and (c) argon (0.53 μm thick).

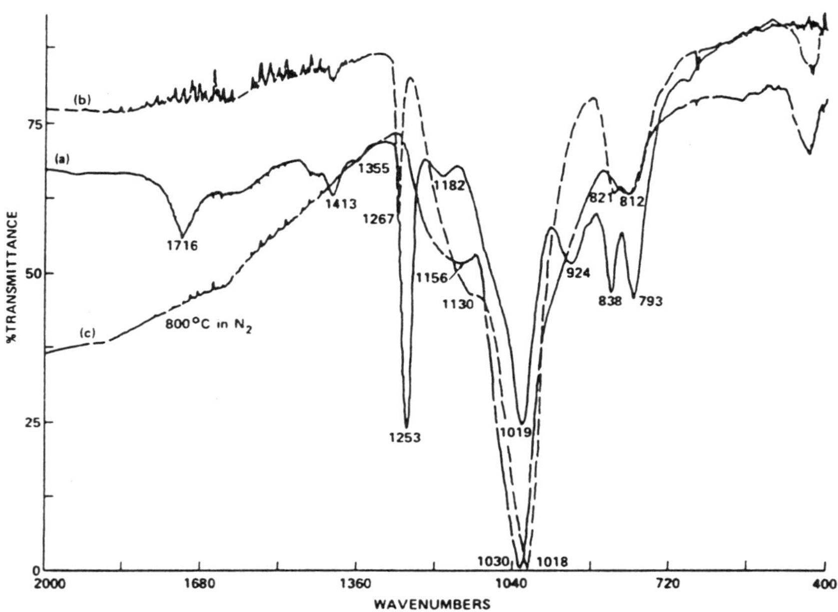


Figure 6. FTIR spectra of PP-HMDS film after different pyrolysis temperature in nitrogen atmosphere: (a) 400°C (b) 600°C and (c) 800°C for 1 hr.

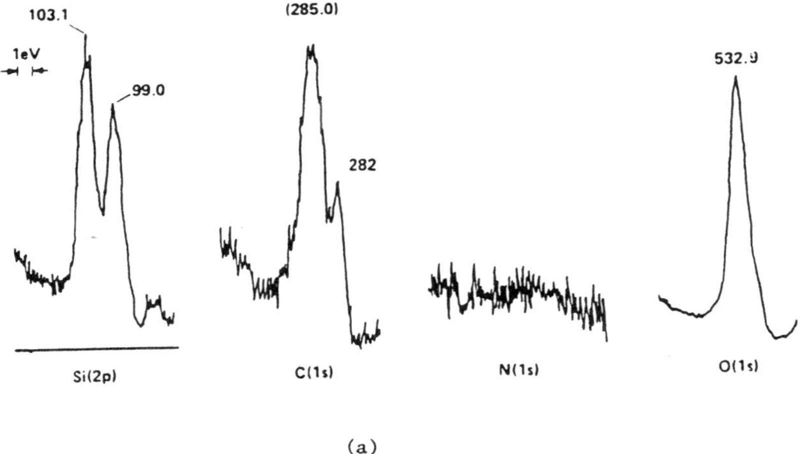


(a)

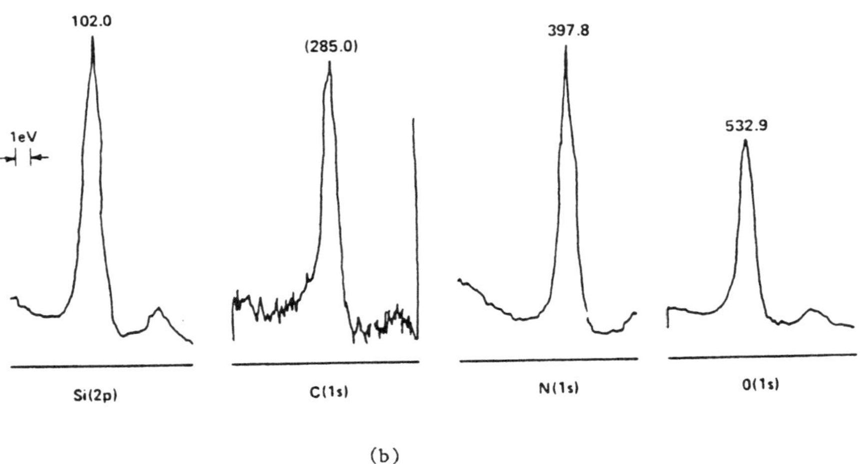


(b)

Figure 7. The ESCA binding energy spectra for commercially available (a) SiC and (b) $Si_3N_4$.

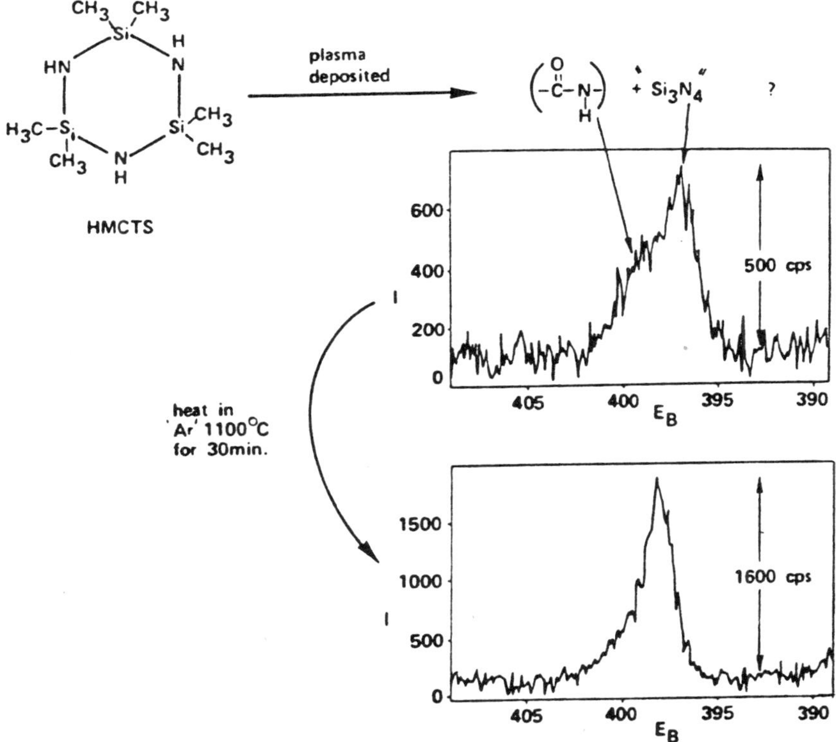


Figure 8. Changes in nitrogen binding energy N(1s), as a function of pyrolysis.

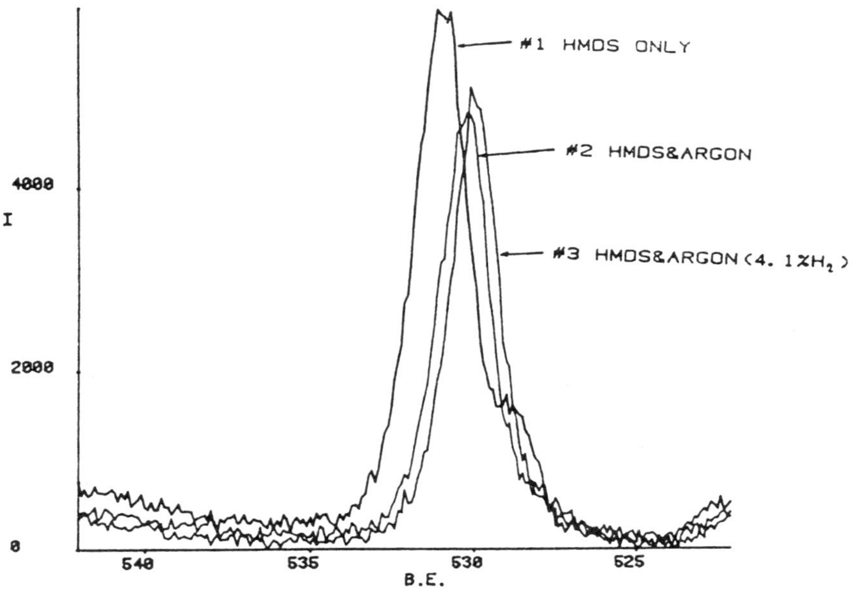


Figure 9.
XPS spectra for oxygen (1s) as a function R.F. plasma polymerization (1) of monomer (2) of monomer after argon purging and (3) of monomer after purging with argon and hydron gas mixture.

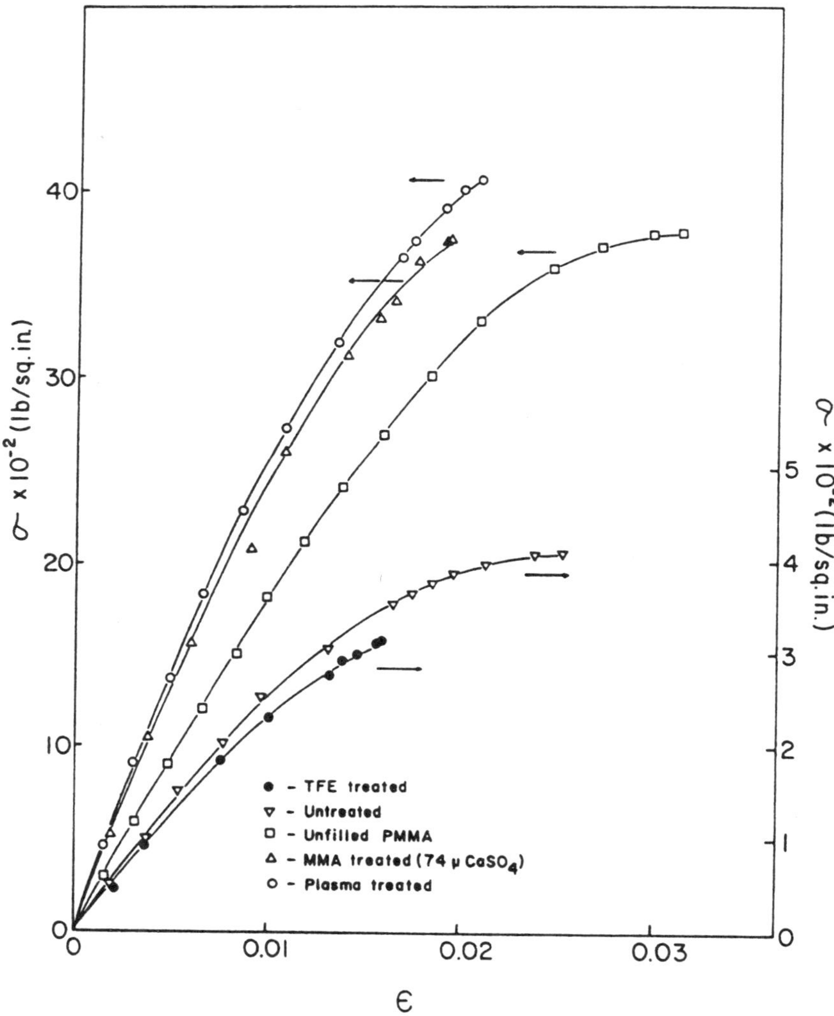


Figure 10. Tensile stress-strain behavior of R.F. plasma coated $CaSO_4$ filler in polymethylmethacrylate. Good bonding methylmethacrylate R.F. plasma treated filler) result in better interfacial adhesion than a poor adhering R.F. plasma "tefla-like" coating on the filler.

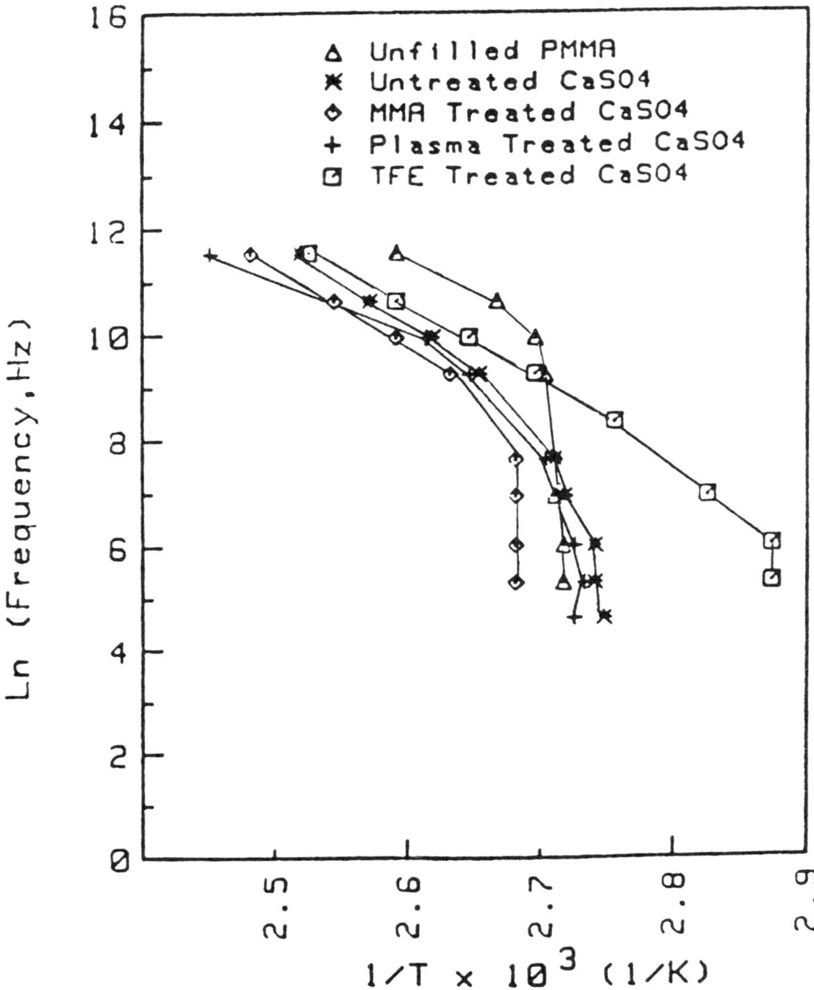


Figure 11. The transition map for the glass transition temperature Tg of poly(methylmethacrylate), PMMA and R. F. plasma tested filler. Note that the good interfacial adhesion coating results in a higher Tg whereas the poor interfacial adhesion coating results in a lower Tg for the composite.

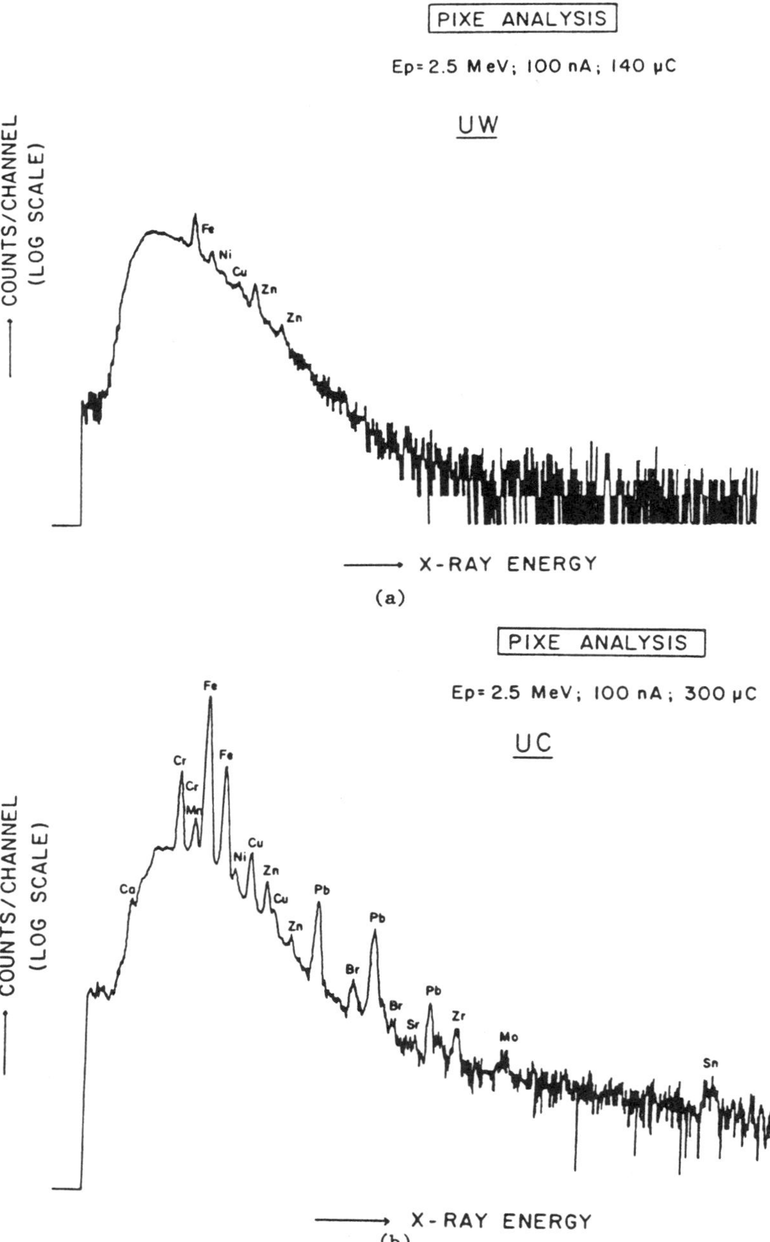


Figure 12.
PIXE spectra for (a) poly(silastyrene) and (b) poly(carbosilane)

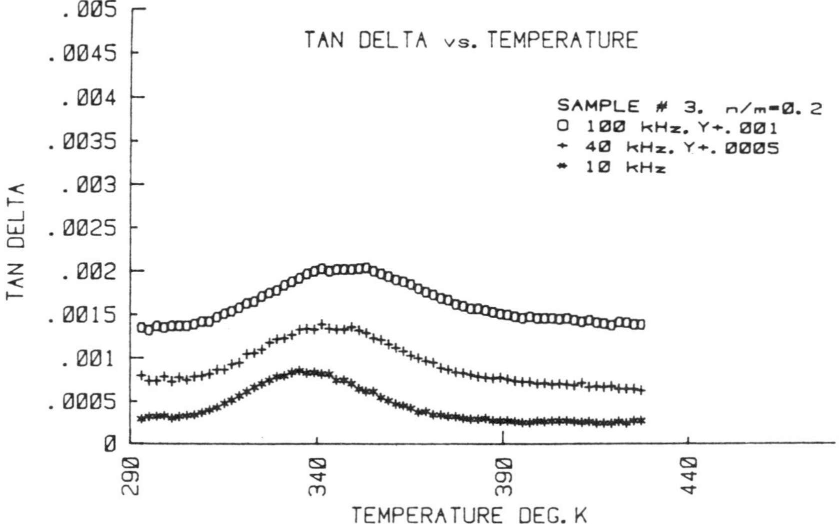


Figure 13. Dielectric tan δ versus temperature for poly(silastyrene) as a function of measurement frequency.

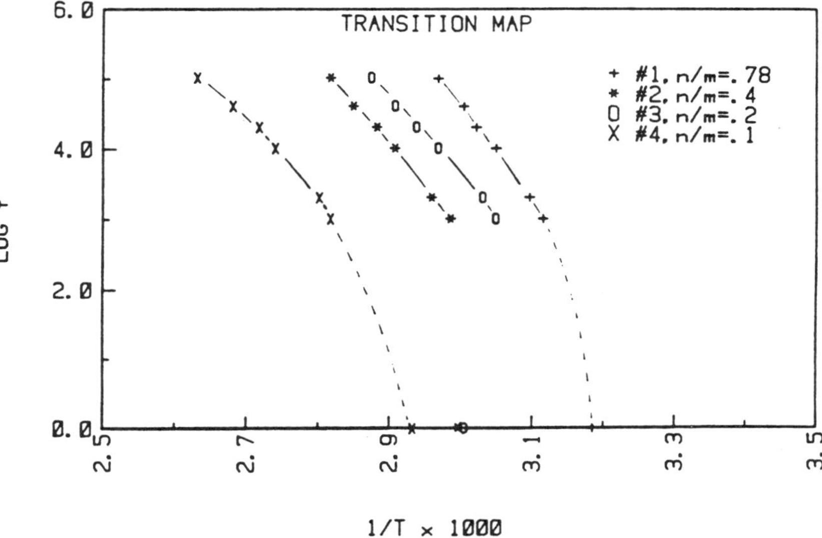


Figure 14. Transition map of the glass transition temperature, Tg, the poly(silastyrene) polymers. The data points at zero measurment frequency, f, are differential scanning calorimetry measurements of Tg.

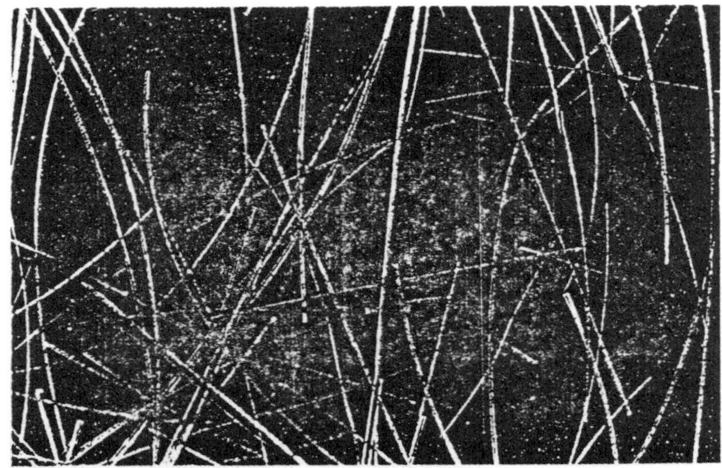

Figure 15(a). Photograph of fibers extruded from poly(silastyrene) sample number one.

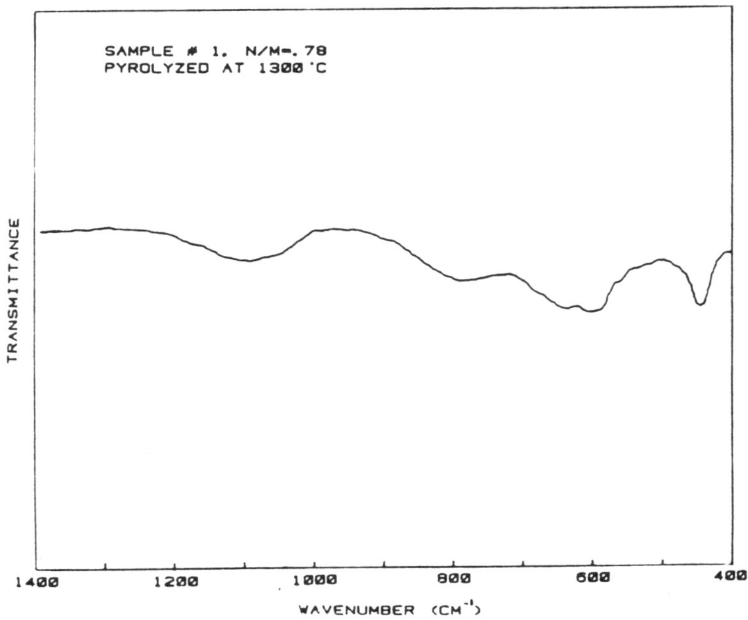


Figure 15(b). Fourier Transform Infrared Spectra of pyrolyzed poly(silastyrene).

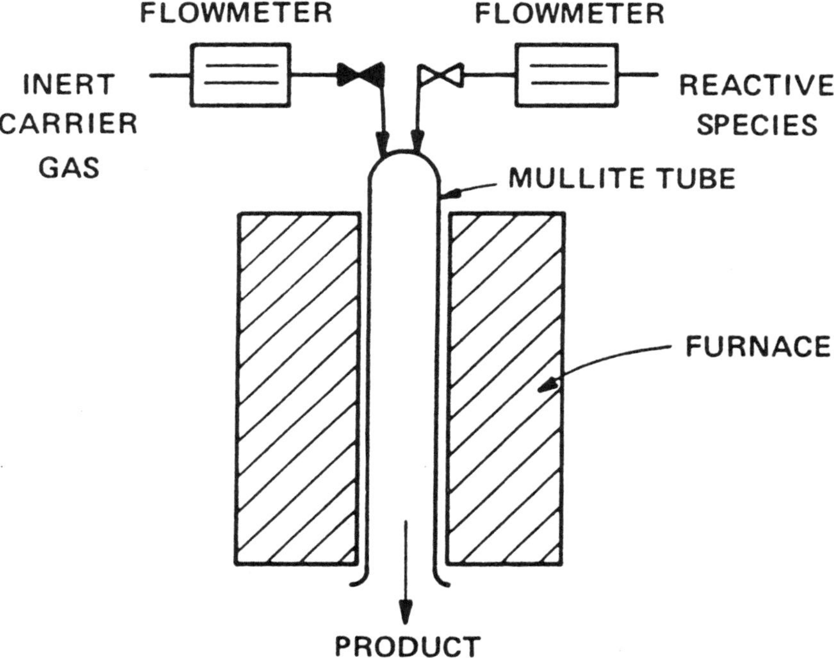


Figure 16. Schematic diagram of the thermal reactor system for producing SiC powder.

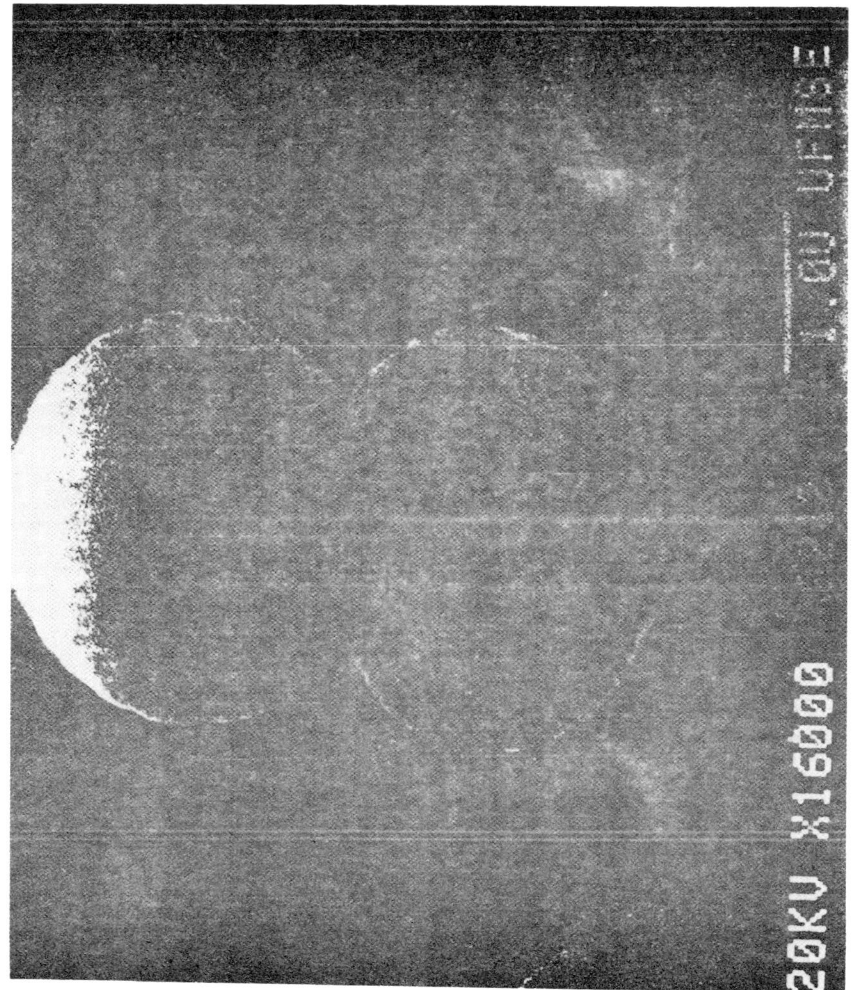


*Figure 17(a). Scanning electron micrograph of large (i.e., ~4μ) SiC particles.*

*Figure 17(b): Scanning electron micrograph of small (i.e., ~0.2 μ) SiC particles.*

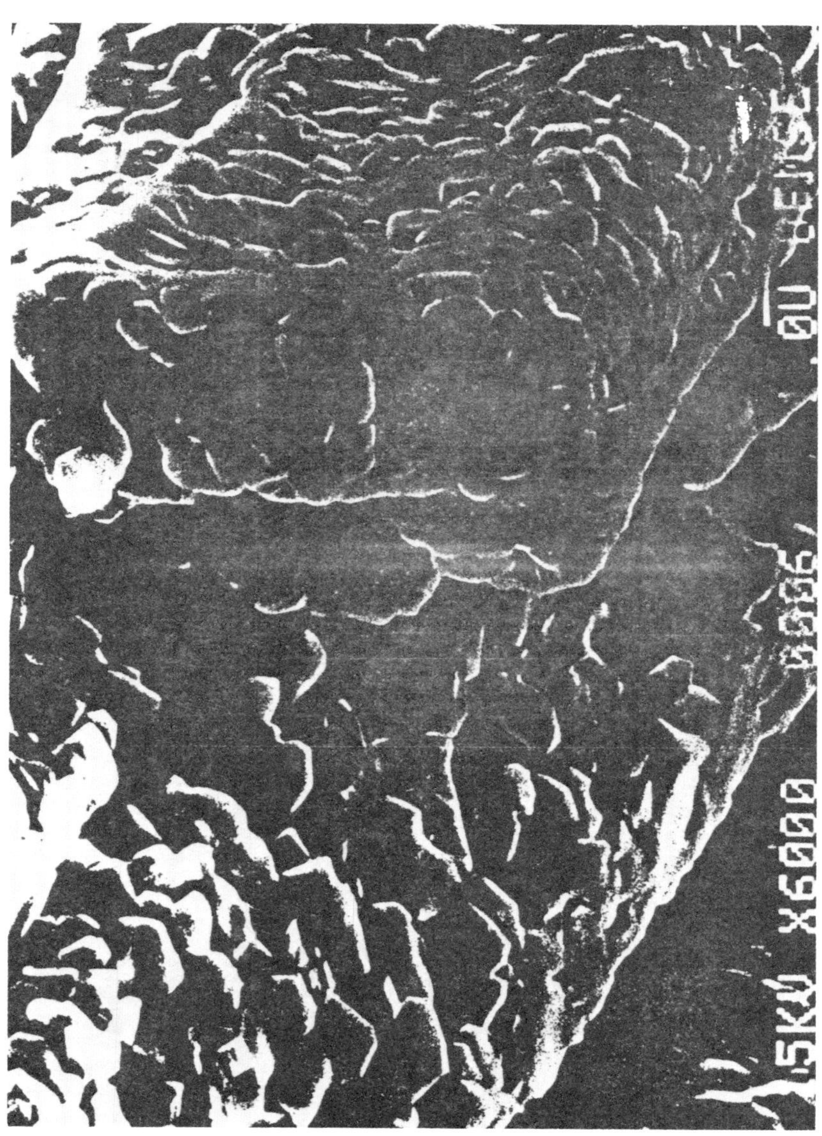

Figure 18. Scanning Electron Micrograph of the $\sim 0.2\mu$ powder (shown in Figure 17b) after compacting at room temperature followed by sintering at 1500°C.

# COMPATIBILITY OF A RANDOM COPOLYMER OF VARYING COMPOSITION WITH EACH HOMOPOLYMER

**K. Fujioka, N. Noethiger, Y. Baba*, A. Kagemoto* and C.L. Beatty**

*Department of Materials Science and Engineering*
*University of Florida*
*Gainesville, FL 32611*

**Department of Chemistry*
*Osaka Institute of Technology*
*Asahi-ku, Osaka 535, Japan*

Abstract

The compatibility between polymers, especially poly(styrene), poly(n-butylmethacrylate), and poly(styrene-co-n-butylmethacrylate) was studied using Visible Spectroscopy, Differential Scanning Calorimetry, Inverse Gas Chromatography and Dielectric Relaxation Spectroscopy. Both the composition of the copolymer and homopolymer with which it was blended and also the percentage of copolymer chains in the blend affected the compatibility. That is, the range of compatibility of the styrene/n-butylmethacrylate random copolymer with polystyrene increases as the styrene content of the copolymer increases. The range of compatibility of styrene/n-butylmethacrylate random copolymers blended with poly(n-butylmethacrylate is greater than that achieved for polystyrene indicating that interactions between n-butylmethacrylate species in the polymer chain are more effective than styrene species in promoting compatibility in polymer blends. Each of the experimental techniques provide slightly different results concerning the degree of compatibility as well as the range of compatibility. These results suggest that a variety of experimental techniques yield are needed to establish the extent and nature of the compatibility of mixed polymers. The dielectric relaxation data suggest that the compositional regions deemed compatible by other techniques and high frequency dielectric measurements appear to be phase separated when examined at low frequencies. These results indicate that the compatibility of polymers may not be limited to strictly compatible or incompatible behavior but may also include intermediate behaviour such as partially mixed regions that coexist with phase separated domains.

## Introduction

The question of compatibility between polymers has been well studied for a number of years by utilizing bulk parameters such as the solubility parameter and the free energy interaction parameter as guides to possible compatibility (1). The lack of success of these approaches for some systems has led to the concept that compatibility may be related to interaction between the specific moieties on the different polymer chains. All of these approaches have had their limitations and have not been universally proven.

However, it is clear that the length of the chain plays a key role, as many oligomers are compatible with polymers of greatly dissimilar chemical composition, yet the high molecular weight polymeric analogues are not compatible (1,2). Therefore, an ideal system for studying the compatibility would be one for which the degree of compatibility would be varied in a continuous fashion and independent of the length of the polymer chain by keeping the length of the polymer chains used to make the blends constant.

An approach that may satisfy the criteria set forth above may be a random copolymer composed of mers of basically incompatible polymers. Thus, polymer chains of a controlled, although variable chemical composition and of a selected molecular weight could be used in blends with the homopolymer formed from each mer or with random copolymers of different mer ratio. Such a random copolymer would be expected to be compatible with homopolymers of one of its mers (3) over some range of composition of the blend. Thus, a limited range of compositional compatibility could be obtained by blending a pure homopolymer chain (A) made up of U mers, (A = (U) = UUU...) with a random copolymer (B) composed of U and V mers, (B = ($V_{1-x}$, $U_x$) = UUVUUV...) where x is a number fraction of U units. In addition, this partial range of compositional compatibility would also be obtained between random copolymers by making the difference between x values sufficiently small. In addition, it may be possible to determine if partial degrees of compatability exist versus strictly compatible or incompatible states in blended polymer systems.

In this paper, poly(styrene) and poly(n-butyl methacrylate) were used as the homopolymers (A and A') and poly(styrene-co-n-butylmethacrylate) of variable chemical composition were used as (B). The compatibility between these polymers [i.e. (A)-(B), (A')-(B) or (B)-(B)] was studied by using visible spectroscopy, Differential Scanning Calorimetry (DSC), Inverse Gas Chromatography (IGC) and Dielectric Relaxation Spectroscopy (DRS).

## EXPERIMENTAL

### Materials

The poly(styrene) (PS) used in this study was purchased from Foster Grant (Cat. No. 8170), which had a number average molecular weight of approximately 100K and a molecular weight distribution ($\bar{M}w/\bar{M}n$) of 2.5. The poly(n-butylmethacrylate) [P(nBMA)] homopolymer was purchased from Polysicences, Inc. and its number average molecular weight was also approximately 100K. Poly(styrene-co-n-butylmethacrylate), [P(S/nBMA)] random copolymers used in this study was purchased from Scientific

Polymer Products, Inc., and also had a number average molecular weight of 100K. The chemical composition ratio (styrene/nBMA) of random copolymers are summarized in Table 1. The molecular weight distribution of all of these polymers was basically unimodal with little skewing of the molecular weight distribution.

Table 1

Chemical Composition of P(S/nBMA)

| denotation | 99/1 | 95/5 | 90/10 | 85/15 | 80/20 | 60/40 | 40/80 | 20/80 | 10/90 | 5/95 |
|---|---|---|---|---|---|---|---|---|---|---|
| styrene wt% | 99 | 95 | 90 | 85 | 80 | 60 | 40 | 20 | 10 | 5 |
| nBMA wt% | 1 | 5 | 10 | 15 | 20 | 40 | 60 | 80 | 90 | 95 |

Visible Spectroscopy

Blending was accomplished by dissolving the polymers in chloroform, which is a good solvent for PS, P(nBMA) and the random copolymer (solubility parameters calculated from the group contribution method (4,5) and heat of vapourization (6) are 9.7, 9.0 and 9.2 at 25°C for PS, P(nBMA), and chloroform, respectively). Films of PS, P(nBMA), P(S/nBMA) and blends between polymers were cast from 5 wt% solutions in chloroform, followed by slow evaporation of the chloroform with protection from contamination by dust. The resulting film was dried under vacuum at room temperature for approximately four days. A Model 552 Perkin-Elmer spectrophotometer was used to measure the visible light transmittance of the cast films. For these measurements, the film thickness was about 0.05 mm for all the polymer films examined.

Differential Scanning Calorimetry (DSC)

Sample preparation was carried out in the same manner as used for Visible Spectroscopy. A Perkin Elmer DSC II was used for measuring the glass transition temperatures of these films. Temperature calibration was achieved using indium, zinc and cyclohexane. Approximately 10 mg (20 mg) portions of the cast film were placed into the aluminum pans and sealed. The heating rate and sensitivity range of DSC were 10°C/min. and 0.5 m cal/s (2 mcal/s) for each measurement, respectively. Values given in parentheses were used for P(nBMA)-P(S/nBMA) system. Samples were heated from 30°C (-73°C) to 160°C (107°C) and then cooled back to 30°C (-73°C) at 160°C/min. After holding for 30 min. at 30°C (-73°C) to achieve equilibrium, the samples were heated to 160°C (107°C) at 10°C/min and immediately cooled back to 30°C (-73°C) at 160°C/min. The values of the reported glass transition temperature, Tg, were determined by this repeated temperature sequence. The glass transition temperature was determined as that temperature corresponding to the midpoint between initial and final baselines of the DSC thermogram.

Inverse Gas Chromatography (IGC)

A Hitachi (Model 163) gas chromatograph equipped with a thermal-conductivity detector was used for the inverse gas chromatography studies. The column temperature was detected by a copper-constantan thermocouple. The average error in column temperature was ± 0.5°C. The flow rate of helium carrier gas was measured by a soap-bubble flowmeter. The column pressure was measured differentially against the atmospheric outlet pressure with a U-tube manometer filled with mercury.

The columns were prepared as reported elsewhere (7,8). The polymers were coated from a benzene (solubility parimeter 9.2 at 25°C) (5) solution onto Chromosorb G (AW-DMCS treated, 70–80 mesh) purchased from Gasukuro Kogyo Co., Ltd. After drying under vacuum for 5 days at room temperature, the coated support was packed into 4 mm I.D. stainless steel columns. The total percent loading of polymer (ca, 12%) on the support was determined by calcination. The relative concentration of polymer in the blends is assumed to be identical with that in the original solution. The columns were conditioned under helium for 5 hours at a temperature above the glass transition temperature of polymers.

Dielectric Relaxation Spectroscopy

The dielectric relaxation cell was purchased from Balsbaugh Company, Inc. A Hewlett Packard (HP) Model 4272A multifrequency digital meter (frequency range $10^2H_z$-$10^5H_z$) was used as the frequency source. This meter has automatic bridge balancing capabilities and was interfaced with a HP 9915A computer for data collection and processing. Plotting of the data was achieved with a HP 9872S plotter.

Dielectric relaxation measurements were made by scanning the temperature at a constant rate of 0.35° C/min in a Delta Design temperature chamber (Model No. 5100) as controlled by the Delta Design Rate Programming Accessory. The temperature of the sample was measured with an accuracy of ±0.1°C by a Fluke 2190A digital thermometer interfaced with the HP 9915A computer that was used for controling the dielectric system and for data acquisition. The samples used in this measurement were prepared in the same manner as described for the visible spectroscopy specimens.

## RESULTS AND DISCUSSION

### Visible Spectroscopy

A common test of compatibility is opacity, since scattering of light occurs from domains of differing refractive index or the lack of light scattering if the domains are approximately less than the wavelength of the radiation used. In this study, visible light of $\gamma$ = 400nm and $\gamma$ = 600nm was used to measure the transmittance of cast films.

The transmittance of cast films are shown in Fig. 1 and 2 for the PS-P(S/nBMA), P(nBMA)-P(S/nBMA) and P(S/nBMA)-P(S/nBMA) systems, respectively. As seen in Fig. 1 for the PS-P(S/nBMA) system, the 99, 95, 90 and 85 wt.% sytrene P(S/nBMA) copolymers are compatible with PS over the entire composition range when blended with PS as detected by the visible spectroscopy. In contrast, the 80, 60, 40 and 20 wt.% styrene P(S/nBMA) copolymers when blended with PS seem to be compatible only in a definite range which becomes smaller as the styrene content in P(S/nBMA) random copolymer decreases.

As seen in Fig. 2 for P(nBMA)-P(S/nBMA) blends the 5 and 10 wt.% styrene copolymer are compatible with P(nBMA) over the entire composition range. However, the 20, 40 and 60 wt% styrene P(S/nBMA) random copolymers seem to be compatible with P(nBMA) over a portion of the composition range which becomes smaller as the styrene wt% of the P(S/nBMA) random copolymer increases. The results for a copolymer/copolymer blend system is also shown in Fig. 2 and the range of composition that exhibits compatibility of blends formed from 40 wt.% styrene P(S/nBMA) and 60 wt.% styrene P(S/nBMA) (i.e., the region of high transparency) is greater than that observed for blends formed from 20 wt.% styrene P(S/nBMA) and 80 wt.% styrene P(S/nBMA) random copolymers. This result is consistent with the hypothesis that as compositional difference between polymers becomes larger the region of compatibility becomes smaller. To quantify the compatibility for PS-P(S/nBMA), P(nBMA)-P(S/nBMA), and P(S/nBMA)-P(S/nBMA) systems mentioned above, the following quantity (defined by eq. 1) was formulated to express the extent or degree of compatibility, as determined by these visible spectrometry measurements. The derived parameter, C, ranges from 0 to 1 with values of 1 being achieved when the blend is optically opaque and a value of 0 representing an optically compatible blend system.

$$C = 1 - T_r = 1 - \frac{2\ Tm}{Ta + Tb} \qquad (1)$$

where $T_a$, $T_b$ and $T_m$ are the transmittance of A polymer, B polymer and the transmittance of a blend of A polymer with B polymer at 50 wt.% of A polymer in blend and $T_r$ is the measured relative transmittance at 50 wt.% for the blends. An equation of this form was selected to aid in the comparison of the visible spectroscopy and the Flory-Huggins interaction parameter results. The C values obtained are plotted against styrene wt.% in P(S/nBMA) random copolymer as illustrated in Figure 3. The compatibility between polymers (Figure 3) could be obtained by making the composition difference between the blended polymers small. In addition, it will be noticed that the composition range for compatibility for the P(nBMA)-P(S/nBMA) system is wider than that for PS-P(S/nBMA) system. The wider composition range for compatibility is thought to be related to the more active or stronger interactions that are possible between n-butmethylacrylate segments than that is possible between styrene repeating units. It is suspected that these interactions are probably of the hydrogen bonding type although that has not been investigated in this work.

Obviously increased interaction between segments in a polymer chain should affect the interaction parameter between chains. The interaction parameter, $\chi_{AB}$ is related to the solubility by the following equation.

$$\chi_{AB} = \frac{V_r}{RT} (\delta_A - \delta_B)^2 \tag{2}$$

where, $V_r$ is the molar volume of repeating unit, $\delta$ is the solubility parameter, R is the molar gas constant and T is the absolute temperature, respectively. The solubility parameter of PS and PnBMA were calculated from the group contribuiton method (4,5) by using the following equation (3).

$$\delta = \rho \Sigma F_i / M \tag{3}$$

$$F_i = (E_i \nu_o)^{1/2} \tag{3'}$$

where $\rho$ is the density of polymer, M is the molecular weight of the repeating unit, $F_i$ is related to cohesive energy ($E_i$) of each functional by eq. (3)' and $\nu_o$ is the molar volume of the group. Equation (4) was used for the calculation of the solubility parameter, $\delta_c$, of the random copolymer, P(S/nBMA).

$$\delta_c = \rho_1 \ (\Sigma F_{i,1}/M_1) \ \phi_1 \ + \rho_2 \ (\Sigma F_{i,2}/M_2) \ \phi_2 \tag{4}$$

where, $\phi_1$ and $\phi_2$ are the volume fraction of each component in the random copolymer. The densities of PS, P(nBMA) required for the calculation were obtained from the literatures (9,10) and the density of P(S/nBMA) was estimated via the role of linear mixtures. The values of $V_r$ and T used for the calculation were 100 $cm^3$/mol and 298K, respectively.

The calculated values of $\chi_{AB}$ are plotted against styrene wt.% in the P(S/nBMA) used in the three blends system described previously in Figure 4. The value of $(\chi_{AB})_{cr}$ (also shown in Figure 4), was estimated as $X_A = X_B = 1000$, by using eq. 5,

$$(\chi_{AB})_{cr} = \frac{1}{2}\left[\frac{1}{X_A{}^{1/2}} + \frac{1}{X_B{}^{1/2}}\right]^2 \qquad (5)$$

In general, if $\chi_{AB} > (\chi_{AB})_{cr}$, then the two polymers should be incompatible and the greater the difference between $\chi_{AB}$ and $(\chi_{Ab})_{cr}$, the more incompatible the polymer should become. If $\chi_{AB} \leq (\chi_{AB})_{cr}$, then the two polymers should be compatible at that composition. Comparing Figure 3 and Figure 4, it seems that the values of C (obtained from visible spectroscopy are in fair agreement with that of $\chi_{AB}$ prediction of compatibility (i.e., for values of $\chi_{AB}$ below $(\chi_{AB})_{cr}$). The similarity of the observed and trends extends to the breath of the composition ranges for blending with PnBMA compared to PS as well as the compatibility of the copolymer/copolymer blends as a function of composition. Thus, it is suggested that the visible spectroscopy may be a convenient and simple method to measure not only the compatibility between polymers but also the degree of incompatibility that is achieved.

Glass Transition Temperature by DSC

To confirm the results of visible spectroscopy, the glass transition temperature of the blends were measured with DSC. The glass transition temperatures obtained were shown in Figure 5 and 6. As seen in Figure 5 in the case of 60, 80 and 85 wt% styrene P(S/nBMA)-PS blends two glass transition temperatures were observed (indicative of phase separation) yet for blends of 90, 95 and 99 wt.% styrene P(S/nBMA) copolymers with PS only one glass transition temperature was observed. This suggests that the range of compatibility is dependent upon the chemical composition of the copolymer as measured by the range of existence of one or two glass transiton temperatures. The 60, 80 and 85 wt.% styrene P(S/nBMA) appear to be compatible with PS in the range of 10-20, 10-15 and 50-60 wt.% PS in the blend, respectively. On the other hand, it appears that 90, 95 and 99 wt.% styrene P(S/nBMA) are totally compatible with PS as determined by DSC measurements.

The single glass transition observed for P(nBMA)-20/80 P(S/nBMA) system (shown in Fig. 6) indicate that these polymers are compatible. Thus, it appears that the composition range for the compatibility for the P(nBMA)-P(S/nBMA) system is wider than that for PS-P(S/nBMA) system analogous to that was found via visible spectroscopy.

Inverse Gas Chromatography (IGC)

The existance of either compatible (i.e., one glass transition temperature) or an incompatible blending (i.e., two glass transition temperatures) dependent upon the chemical composition of the random copolymer used in the blend suggests that the interaction parameter between the random copolymer and the homopolymer is a function of the chemical composition of the random copolymer. Thus, inverse gas phase chromatography has been utilized to determine the interaction parameter, $\chi'_{23}$, of the random copolymer and the homopolymer in the PS-80/20 P(S/nBMA) and PS-60/40 P(S/nBMA) blend systems.

Specific retention volume, $V^o_g$ ($cm^3/g$) was computed from the familiar method (11). Flory-Huggins interaction parameters, $\chi$, (12) at infinite dilution of the probe were determined in the usual manner (13) from the measured specific retention volume for benzene in each of the pure phases (i.e., $\chi_{12}$ and $\chi_{13}$) (13) as well as for the mixed stationary phase ($\chi_{1(23)}$) (14) using eq. (6), (6)' and (7).

$$\chi_{12} = \ln \left[\frac{RT\nu_2}{V^o_g V_1 P_1^0}\right] - \left(1 - \frac{V_1}{\bar{M}_2 \nu_2}\right) - \frac{P_1^o}{RT}(B_{11} - V_1) \quad (6)$$

$$\chi_{13} = \ln \left[\frac{RT\nu_3}{V^o_g V_1 P_1^o}\right] - \left(1 - \frac{V}{\bar{M}_3\ 3}\right) - \frac{P_1^o}{RT}(B_{11} - V_1) \quad (6)'$$

$$\chi_{1(23)} = \chi_{12}\phi_2 + \chi_{13}\phi_3 - \chi'_{23}\phi_2\phi_3 = \ln \left[\frac{RT(\omega_2\nu_2 + \omega_3\nu_3)}{P_1^o V^o_g V_1}\right] -$$

$$\left(1 - \frac{V_1}{\bar{M}_2\nu_2}\phi_2 - \frac{V_1}{\bar{M}_3\nu_3}\phi_3\right) - \frac{P^o_1}{RT}(B_{11} - V_1) \quad (7)$$

where, R: gas constant, T: temperature in K, $V^o_g$: specific retention volume, $\nu_2$, $\nu_3$; specific volume of polymer (2) and polymer (3), $P_1^\circ$: vapour pressure of pure solvent, $V_1$ = molar volume of solvent, $B_{11}$: second virial coeffecient of solvent, $\bar{M}_2$, $\bar{M}_3$: molecular weight of polymers, $\omega_2$, $\omega_3$: weight fraction of polymers in mixed stationary phase.

The $\chi'_{23}$ values for the PS-80/20 P(S/nBMA) and the 60/40 P(S/nBMA) blend systems are shown in Figure 7. In the PS-80/20 P(S/nBMA) system, for blend compositions less than 90 wt.% PS wt. fraction, the $\chi_{23}'$ value are positive or zero, (i.e., incompatibility should be found) whereas in the range of more than 90 wt.% PS the $\chi'_{23}$ value are negative, (i.e., these systems should be compatible). In the PS 60/40 P(S/nBMA) system the value of $\chi'_{23}$ versus PS wt.% depends on the temperature of measurement. The results obtained at 393°K and 413°K indicate that compatibility can be achived at greater than 85 wt% PS in the PS-60/40 P(S/nBMA) blend. However, the result at 445°K shows that these two polymers are incompatible over the entire composition range. It is suspected that at 445°K the lower critical solution temperature (LCST) has been exceeded. These results suggest that inverse gas phase chromatography can be used to study critical solution temperatures (CST) (15,16).

Dielectric Relaxation Spectroscopy

Dielectric relaxation spectroscopic measurements were made on the PS-80/20 P(S/nBMA) blend system as a function of blend composition to determine if the other techniques utilized (i.e., vs. DSC, and IGC) were sufficiently sensitive to interactions at the segmental and molecular level. The effect of frequency on the dielectric tan $\delta$ versus temperature plots is presented for this blend system in Figure 8(a) and 8(b) which show the results of PS homopolymer and 80/20 P(S/nBMA), random copolymer, respectively. Figure 8(c) and 8(d) show the results of blends, 80 wt.%, PS-80/20 P(S/nBMA) and 60 wt.% PS-80/20 P(S/nBMA) blends, respectively. As seen in Figure 8a, the shift of the glass transition temperature appears to shift to higher temperatures as the measurement frequency is increased as was expected for all the homopolymers and random copolymers. In all of the blends which have the blend composition of 40 at 70 PS or greater in the blend, there is only one apparent glass transition temperature at high frequencies (e.g., 100 kHz). However, in nearly all cases (i.e., ≤90 wt.% PS) the blends exhibit two peaks at low frequencies (i.e., ≤400 Hz). Clearly this low frequency dielectric relaxation data suggests that phase separation occurs at the molecular level with the merger of the two glass transition peaks at the higher measurement frequencies. These results also clearly indicate that the detection of phase separation depends upon the sensitivity and volume of material that is probed by the particular technique that is utilized. Thus, it appears that, of the techniques utilized in this work that the dielectric relaxation data at high frequencies agrees better with the calorimetric and visible spectroscopy data in terms of compatible vs. incompatible behavior whereas the low frequency data allows detection of phase separation at the segmental or molecular level even in the blend composition ranges that appear to be compatible via the other two techniques.

To better understand the affect of measurement frequency or the detection of one or two phases (i.e., Tg's), transition maps were constructed for this series of blends of the homopolymer polystyrene blended with the 80 wt.% styrene/20 wt.% n-butylmethacrylate random copolymer. The data in Fig. 9a indicates that both PS and the 80/20 P(S/nBMA) random copolymer exhibit typical Williams-Landel-Ferry type behavior for their glass transition relaxation process. This data is shown on the succeeding transition maps for the blends as dashed lines as reference lines to facilitate comparison of the blend data with the homopolymer behavior. The 90 wt.% PS homopolymer-10 wt.% P(S/nBMA) random copolymer blend transition map data in Figure 9b illustrates that (1) the high frequency single peak almost perfectly coincides with the PS homopolymer data and (2) the high frequency copolymer peak is not observable. However, the low frequency data does not coincide well with the previously obtained data for the unblended polymers. The data suggests that the polystyrene rich phase, or the polystyrene-like relaxation process, may be anti-plasticized whereas the random copolymer-rich phase, or the random copolymer-like relaxation process, may be plasticized. The transition map for the 80 wt.% PS/20% (80/20) P(S/nBMA) random copolymer blend (Figure 9c) illustrates different behavior in that the highest frequency glass transition temperature peak is depressed to values below the 80/20 P(S/nBMA) random copolymer glass transition temperature which suggests that plasticization may be evident even in the high frequency measurement regime. In addition, two relaxation process that are observed at frequencies below 100kHz both occur at lower temperatures than the relaxation process that are observed for the unblended polymers. In Figures 9d and 9c it becomes apparent that when data is taken at closer frequency intervals a transition from a two peak to a single merged apparent relaxation process can be observed. In these data as well as in Figure 9f the blends formed are incompatible as measured by both the calorimetric and visible spectroscopic techniques. The transition map data is summarized in Figure 10a and 10b. It is interesting to note that the relaxation process that occurs at high temperature in the low frequency regime for all of these blend compositions (the upper set of curves in Figure 10b) is nearly identical with the PS homopolymer data (i.e., 100 % PS) suggesting that little mixing occurs at the phase boundary and that the polystyrene rich phase is quite incompatible. In contrast, the low frequency data for the random copolymer rich phase occurs at a lower temperature for all the blend compositions at 200 Hz. This suggests as mentioned previously that plasticization may be occuring. However, plasticization by a higher glass transition polymer such as polystyrene seems unlikely. A more likely plasticizing species would be water (as PnBMA is hydroscopic) but all of blended as well as the polystyrene and random copolymer species that were blended were carefully vacuum dried and handled identically. Thus, the mechanism for a reduction of the temperature of this blend relaxation process is unresolved.

The high frequency (single relaxation process) glass transition temperature appears to gradually decrease in the high weight percent polystyrene ($\geq$ 90 wt.%) blend region. This result would be expected for a compatible blend system. The composition range over which this behavior occurs corresponds rather well with the range of compatibility of these two polymers as determined by both the calorimetry and visible spectroscopy. In the composition range where incompatibility is observed, by DSC and VS, the high frequency data indicate that the combined relaxation process occurs at a lower temperature than the PS glass transition temperature. The temperature of this combined relaxation process appears to pass through a minimum in the 60–80 wt.% PS for this blend pair with a gradual to the copolymer Tg as the PS content is decreased in the blend system. This minimum in the high frequency data (i.e., 40 KHz) corresponds roughly to the copolymer rich phase in the low frequency (i.e., 200 Hz) data. These minima indicate that although the two polymers appear to phase separate there may be significant mixing occuring at the phase boundary near the 70 wt.% PS blend composition. The results on this blend system to date suggest that similar additional research on polymer blends by a variety of investigative methods is warranted.

Summary

1) Visible Spectroscopy, VS, allows the determination of the concentration range for which a random copolymer remains transparent when mixed with a homopolymer of one mer from the copolymer.

2) Differential Scanning Calorimetry (DSC) and Inverse Gas Chromatography (IGC) suggest that the concentration range of compatibility may be smaller than that determined by visible spectroscopy but does exist.

3) Dielectric Relaxation Spectroscopy Spectroscopic data indicate that the range of compatibility is smaller yet (if it exist at all) than the range of compatibility detected by VS and DSC. The detectability of two phases (i.e., two Tg's) is dependent upon the frequency of measurement.

4) At concentrations less than 10 wt.% of the 80/20 P(S/nBMA) random copolymer blended with PS, the high frequency Tg approaches the value for PS. At higher concentrations of the P(S/nBMA), copolymer, the Tg of the blend is lower than that of P(S/nBMA) copolymer at all frequencies of measurement.

5) All of these results suggest that a variety of experimental techniques give more insight into the extent and nature of the compatibility of polymers. Further research will be needed to examine the interaction between specific moieties on the polymer chain via FTIR, ESCA, etc.

## Acknowledgments

This work was partially supported by the Microstructure Center of Excellence and the Air Force Office of Scientific Research; Contract # F49620-80-C-0047.

## References

1. D. R. Paul and S. Newman, "Polymer Blends", Academic Press, p. 115 N.Y. (1978). O. Olabisi, L. M. Robeson and M. T. Shaw, "Polymer-Polymer Miscibility", p. 19 and p. 117, N.Y. (1979).

2. G. DiPaola-Baranyi, Polymer Preprints, 21, No. 2 214 (1980).

3. P. de Gennes, "Scaling Concepts in Polymer Physics", Cornell Univ. Press., Ithaca and London, p. 100 (1979).

4. P. A. Small, J. Appl. Chem., 3, 71 (1953).

5. D. W. Van Krevelen "Properties of Polymers", 2nd Ed., p. 134 Elsevier, Amsterdam (1976).

6. J. Brandrup and E. H. Immergut, "Polymer Handbook", 2nd Ed. IV 337 (1975).

7. G. DiPaola-Baranyi, Macromolecules, 14, 683 (1981).

8. G. DiPaola-Baranyi, Macromolecules, 14, 1456 (1981).

9. O. Olabisi and R. Simha, Macromolecules, 8, 206 (1975).

10. H. Hocker, G. J. Blake and P. J. Flory, Trans Faraday Soc., 67, 2251 (1971).

11. A. B. Littlewood, C.S.A. Phillips and D. T. Price, J. Chem. Soc. 1480 (1955).

12. P. J. Flory, "Principles of Polymer Chemistry"; Cornell University, Press: Ithaca, N.Y., (1953).

13. D. Patterson, Y. B. Tewari, H. P. Schreiber and J. E. Guillet, Macromolecules, 4, 356 (1971).

14. D. D. Deshpande, D. Patterson, H. P. Schreiber and C. S. Su, Macromolecules, 7, 530 (1974).

15. G. Delmas and D. Patterson, J. Paint Technol., 34, 677 (1962).

16. T. K. Kwei, T. Nishi and R. F. Robert, Macromolecules, 7, 667 (1974).

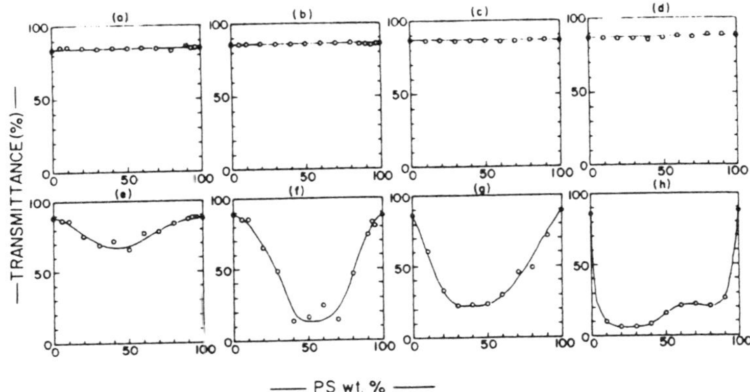


Fig. 1 Transmittance at λ = 400 nm of PS blended with P(S/nBMA) random copolymers having various chemical composition: (a) 99/1, (b) 95/5, (c) 90/10, (d) 85/15, (e) 80/20, (f) 60/40, (g) 40/60 and (h) 20/80.

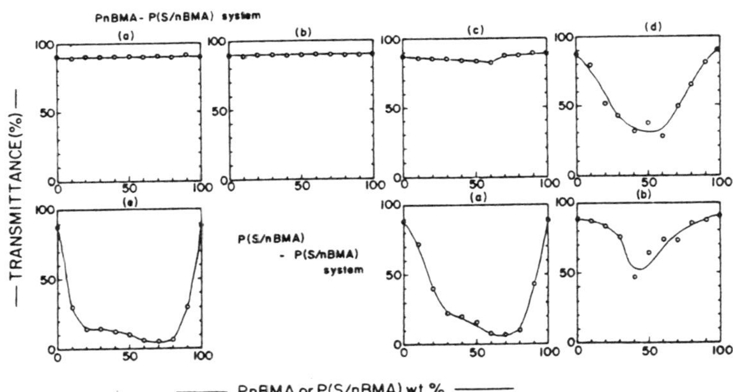


Fig. 2 Transmittance at λ = 400 nm of P(nBMA) blended with P(S/nBMA) random copolymers having various chemical composition (a) 5/95, (b) 10/90, (c) 20/80, (d) 40/60, transmittance at λ = 400 nm for P(S/nBMA) P(S/nBMA) random copolymers (a) 20/80-80/20, (b) 40/60-60/40.

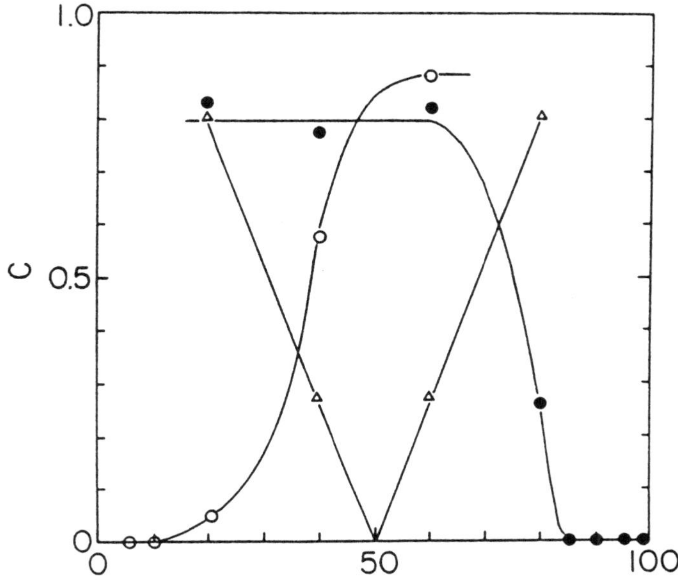


Fig. 3 Dependence of C on PS wt.% various blend systems.
(a): PS-P(S/nBMA) system
(b): P(nBMA)-P(S/nBMA) system
(c): P(S/nBMA)-P(S/nBMA) system.

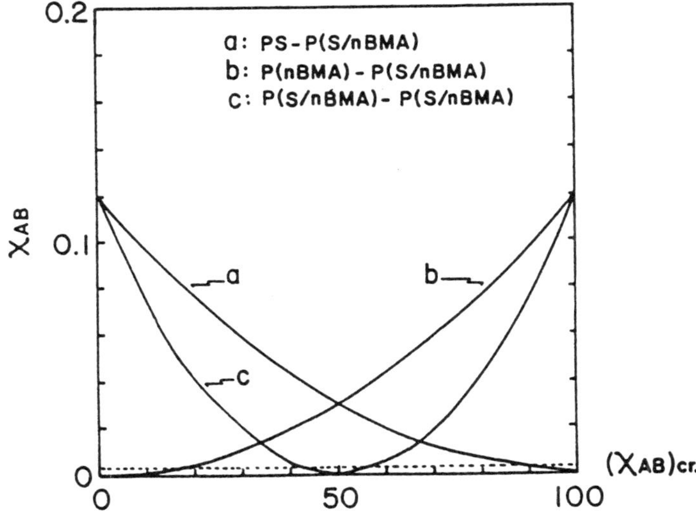


Fig. 4 $\chi_{AB}$ calculated from the group functional method for styrene wt.% in PS-P(S/nBMA) blends.

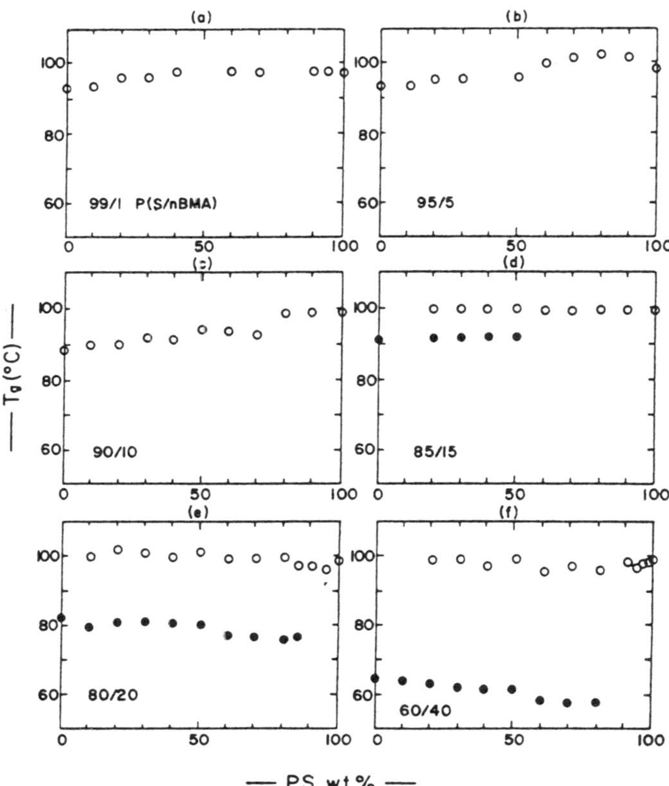


Fig. 5 Glass transition temperature of PS blended with P(S/nBMA) random copolymers having various chemical composition (a) 99/1, (b) 75/5, (c) 90/10, (d) 85/15, (e) 80/20 and (f) 60/40.

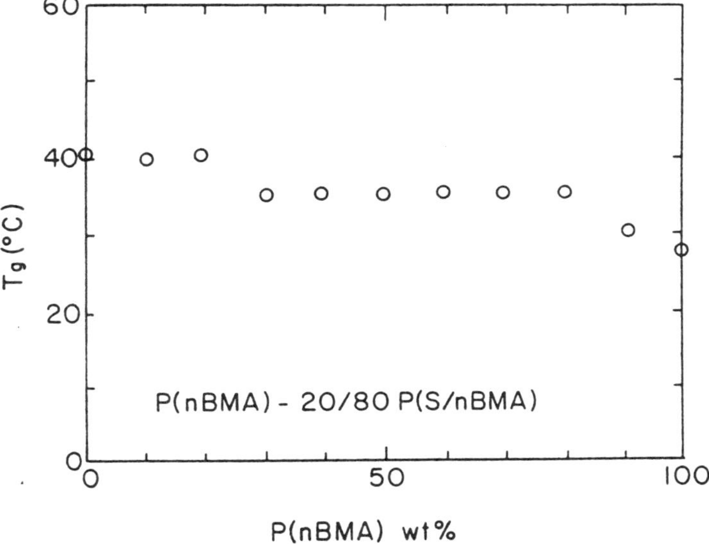


Fig. 6 Glass transition temperature of P(nBMA) blended with 20/80 P(S/nBMA) as detected by Differential Scanning Calorimetry.

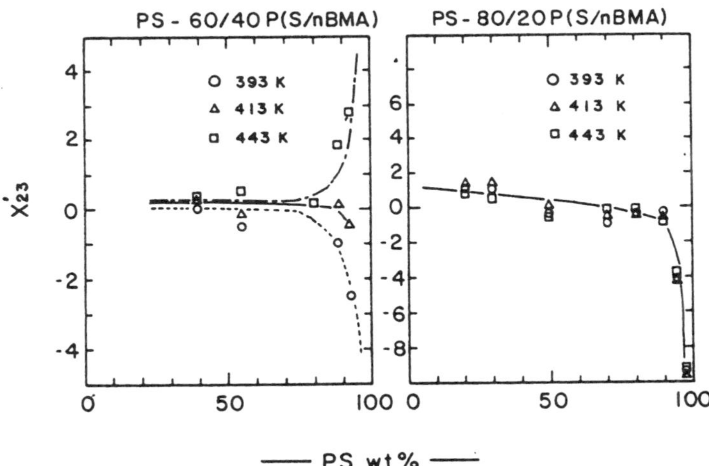


Fig. 7 Free energy interaction parameter $\chi_{23}$ between polymers for PS-60/40 P(S/nBMA) and PS-80/20 P(S/nBMA) blend systems.

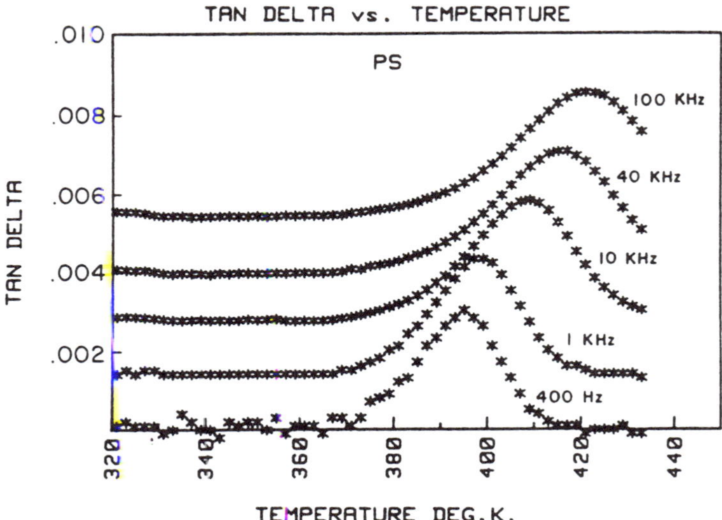


Fig. 8a Dielectric tan δ of PS versus temperature as a function of measurement frequency.

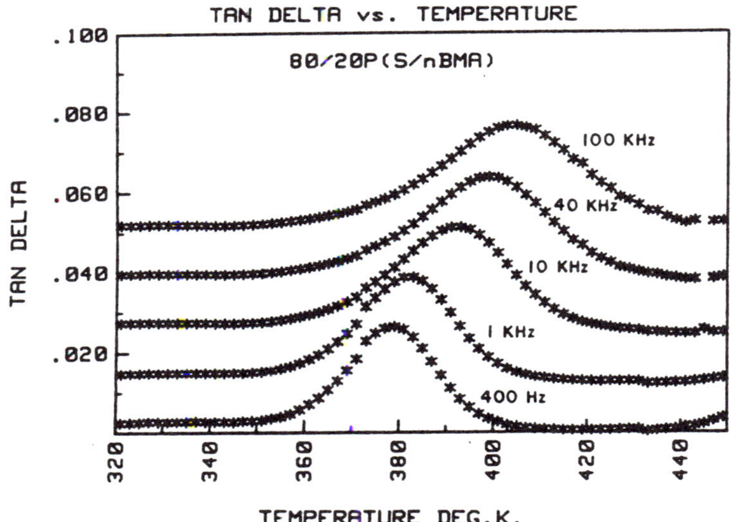


Fig. 8b Dielectric tan δ of 80/20 P(S/nBMA) versus temperature as a function of measurement frequency.

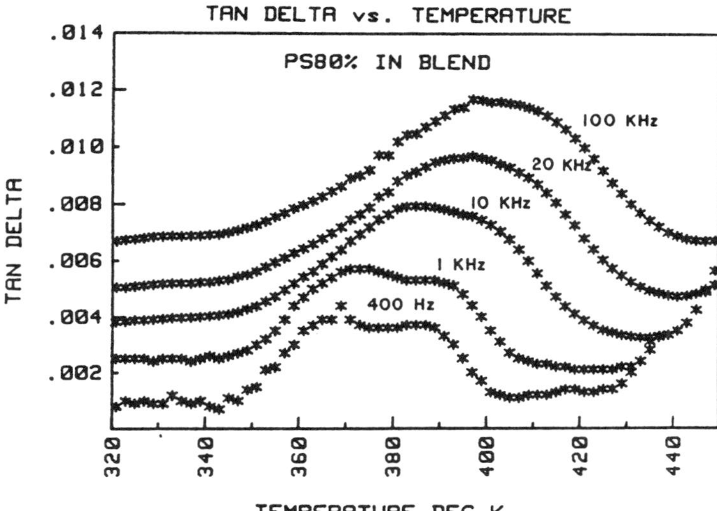


Fig. 8c Tan δ of PS blended with 80/20 P(S/nBMA)[PS 80%, 80/20 P(S/nBMA) 20%] versus temperature as a function of measurement frequency.

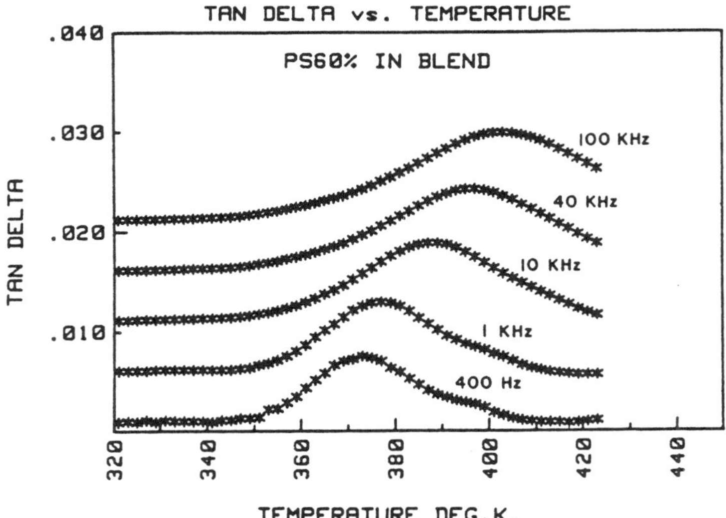


Fig. 8d Tan δ of PS blended with 80/20 P(S/nBMA) [PS 60%, 80/20 P(S/nBMA) 40% versus temperature as a function of measurement frequency.

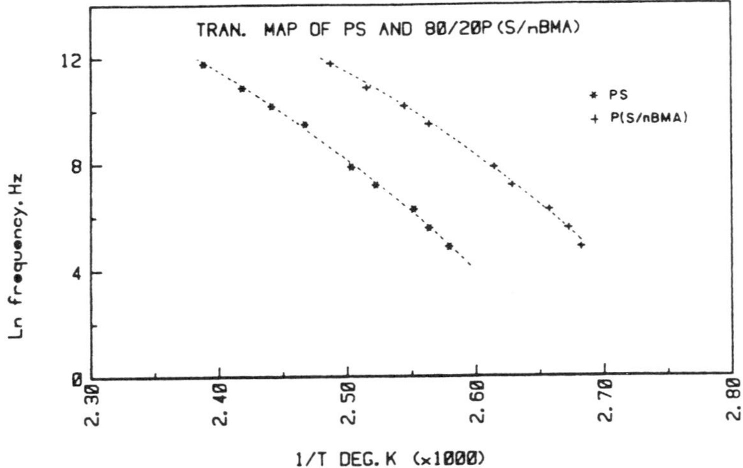


Fig. 9a Glass transition temperature transition map of PS and 80/20 P(S/nBMA) homopolymers.

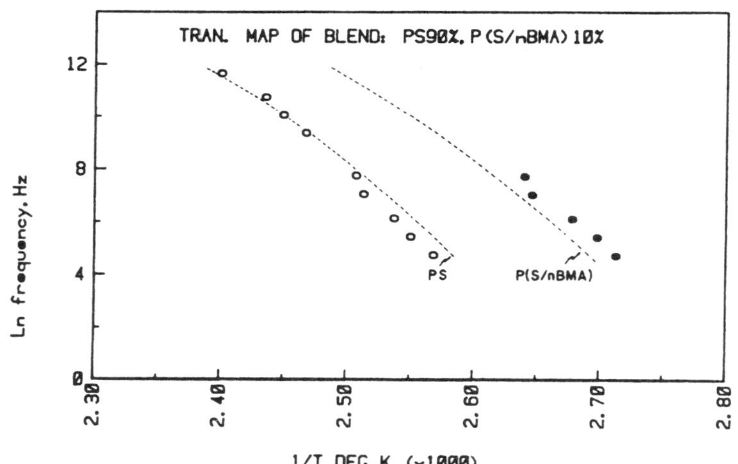


Fig. 9b Glass transition temperature transition map of the blend: PS 90% 80/20 P(S/nBMA) 20%.

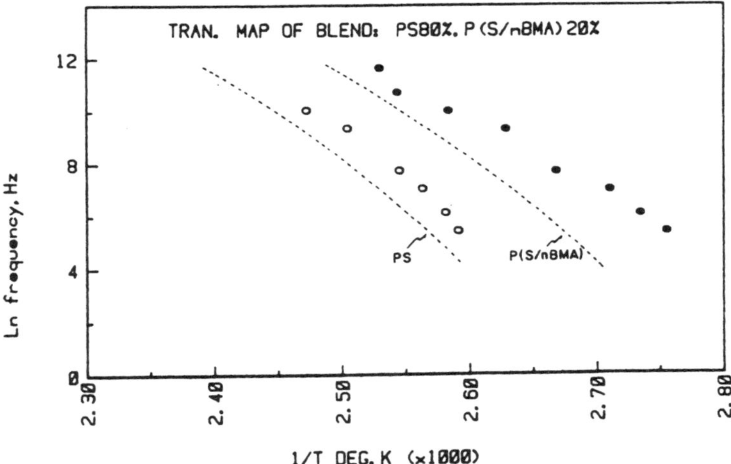


Fig. 9c Transition map of the blend: PS 80%, 80/20 P(S/nBMA) 20%.

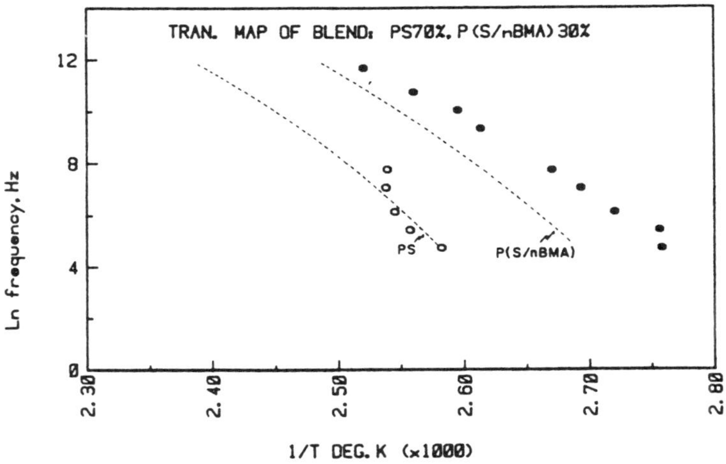


Fig. 9d Transition map of the blend: PS 70%, 80/20 P(S/nBMA) 30%.

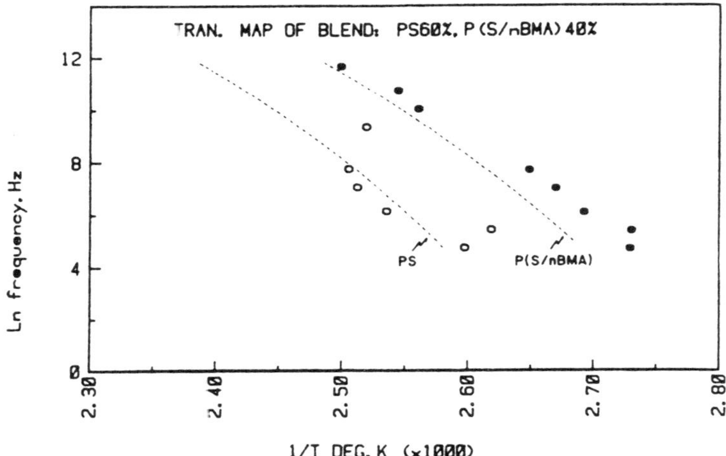


Fig. 9e Transition map of the blend: PS 60%, 80/20 P(S/nBMA) 40%.

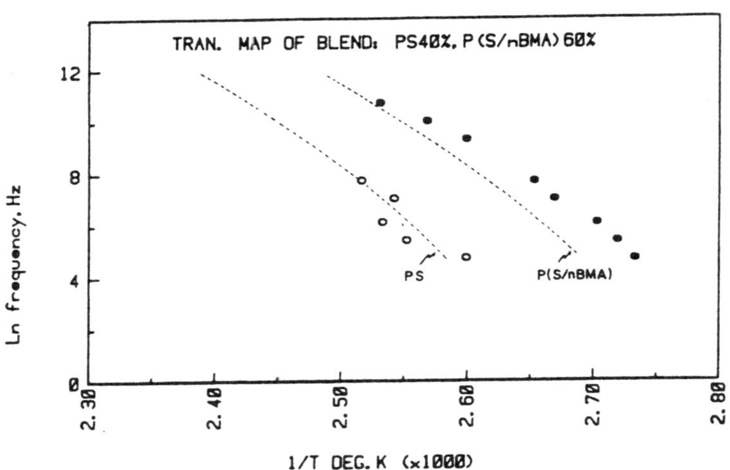


Fig. 9f Transition map of the blend: PS 40%, 80/20 P(S/nBMA) 60%.

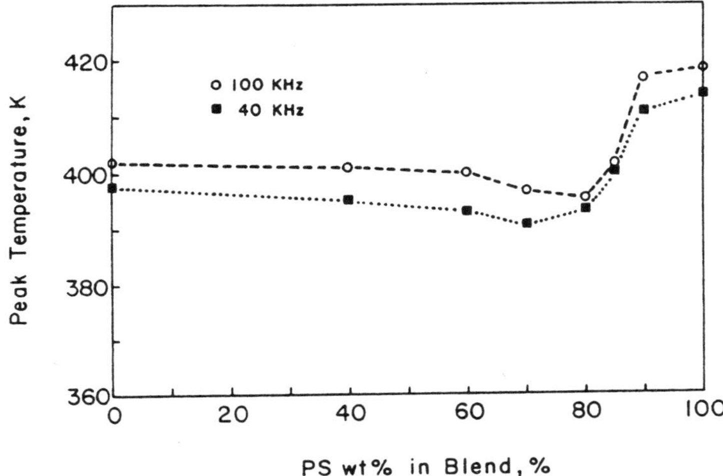


Fig. 10a Dependence of transition temperature on PS wt.% in blend at high dielectric relaxation measurement frequencies.

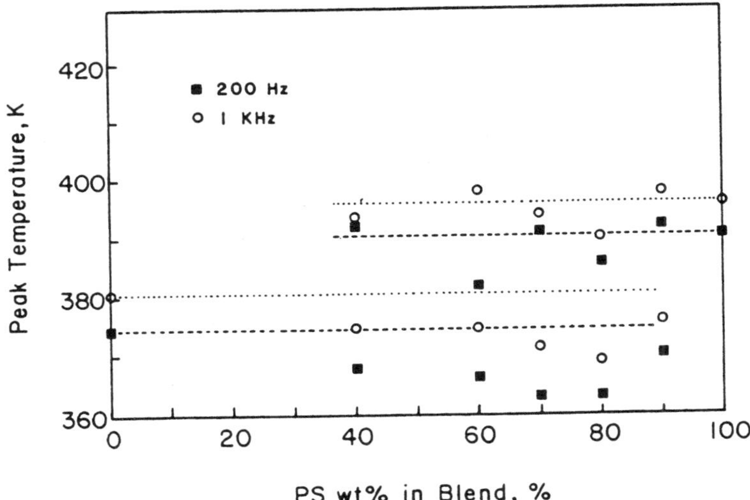


Fig. 10b Dependence of peak temperatures on PS wt.% in blend at low measurement frequencies. The dashed and dotted lines represent the transition temperatures observed at the specified measurement frequency for the PS and P(nBMA) homopolymers.

# LIST OF PUBLICATIONS AND PRESENTATIONS EITHER PARTIALLY OR FULLY SUPPORTED BY AFOSR CONTRACT (1982-83)

1. "Treatment of Glass Surfaces for Polymer-Glass Composites," C. T. Lee, D. E. Clark, K. Shih and C. L. Beatty, presented at the 7th Annual Conference on Composites and Advanced Materials, Jan. 16-19, 1983, Cocoa Beach, FL.

2. "Phase Transformations in Sol-Gel Derived Aluminas," D. E. Clark and J. J. Lannutti, presented at the International Conference on Ultrastructure Processing of Ceramics, Glasses and Composites, February 13-17, 1983, Gainesville, FL, L. L. Hench and D. R. Ulrich, eds., J. Wiley & Sons (1983).

3. "Static Aqueous Corrosion of Soda-Disilicate Glass (33N) in Simulated Geologic Environments," Y. Chao, M. Taie and D. E. Clark, presented at Annual Meeting of the American Ceramic Society, May 5, 1983, Cincinnati, OH.

4. "Weathering of Binary Alkali Silicate Glasses and Glass-Ceramics," Y. Chao and D. E. Clark, in Ceramic Engineering and Science Proceedings, 3[9-10] 458-476 (1982).

5. "Surface Analysis of Glasses," L. L. Hench and D. E. Clark, in Industrial Applications of Surface Analysis, L. A. Casper and C. J. Powell, eds., ACS Symposium Series, ISSN 009-6156, 199 (1982) pp. 203-229.

6. "Surface Behavior of Gel-Derived Glasses," L. L. Hench, M. Prassas, and J. Phalippou, in Ceramic Engineering and Science Proceedings, 3[9-10] 477-483 (1982).

7. "Low Temperature Oxidation of SiC," Bulent O. Yavuz and L. L. Hench, in Ceramic Science and Engineering Proceedings, 3[9-10] 596-600 (1982).

8. "Glass Surfaces - 1982," L. L. Hench in Proceedings of International Conference on Physics of Amorphous Solids, J. Zarzycki, ed., Montpellier, France, July 1982.

9. "Preparation of $xNa_2O-(1-x)SiO_2$ Gels for the Gel-Glass Process. Part I: Atmospheric Effect on the Structural Evolution of the Gels," M. Prassas, L. L. Hench, J. Phalippou and J. Zarzycki, J. Non-Cryst. Solids, 48, 79-95 (1982).

10. "Preparation of $xNa_2O-(1-x)SiO_2$ Gels for the Gel-Glass Process. Part II: The Gel-Glass Conversion," M. Prassas, J. Phalippou and L. L. Hench, J. Non-Cryst. Solids (in press).

11. "Physical Chemical Factors in Sol-Gel Processing," M. Prassas and L. L. Hench, in Ultrastructure Processing of Ceramics, Glasses and Composites, L. L. Hench and D. R. Ulrich, eds., J. Wiley & Sons (1983).

12. "Processing and Environmental Behavior of a 20 Mol % $Na_2O$ - 80 Mol % (20N) $SiO_2$ Gel-Glass," L. L. Hench, S. Wallace, S. H. Wang and M. Prassas, presented at 7th Annual Conference on Composites and Advanced Ceramic Materials, Jan. 16-19, 1983, Cocoa Beach, FL.

13. "Aqueous Durability of Lithium Disilicate Glass-Ceramics," W. J. McCracken, D. E. Clark, and L. L. Hench, Bull. Am. Ceram. Soc., 61[11] 1218-1223 (1982).

14. "Physical Parameters for Characterizing Agglomerates," G. Y. Onoda and R. T. DeHoff, presented at the Annual Meeting of the American Ceramic Society, May 5, 1982, Cincinnati, OH.

15. "Role of Effective Stress in Extrusion," M. A. Janney and G. Y. Onoda, presented at the Annual Meeting of the American Ceramic Society, May 5, 1982, Cincinnati, OH.

16. "Application of Specific Volume Diagrams to Ceramic Processing," G. Y. Onoda, presented at the Annual Meeting of the American Ceramic Society, May 5, 1982, Cincinnati, OH.

17. "Surface Chemistry of Oxides in Water," G. Y. Onoda, presented at the International Conference on Ultrastructure Processing of Ceramics, Glasses and Composites, Feb. 1983, Gainesville, FL.

18. "Application of Soil Mechanics Concepts to Ceramic Particulate Processing," in Advances in Powder Technology, G. Chin, ed., American Society for Metals, 1982.

19. Specific Volume Diagrams for Ceramic Processing," G. Y. Onoda, Jr., to be published in J. American Ceramic Society, April 1983.

20. "Surface Chemistry in Water, G. Y. Onoda and J. A. Casey, in Ultrastructure Processing of Ceramics, Glasses and Composites, L. L. Hench and D. R. Ulrich, eds., J. Wiley & Sons (1983).

21. "$Si_3N_4$ and SiC from Organometallic Precursors," C. L. Beatty, in Ultrastructure Processing of Ceramics, Glasses and Composites, L. L. Hench and D. R. Ulrich, eds., J. Wiley & Sons (1983).

22. "Compatibility of a Random Copolymer with Each Homopolymer," K. Fujioka, N. Noethiger, C. L. Beatty, A. Kugemoto and Y. Baba, presented at the ACS/AIChE joint meeting on Polymer Compatibility, Nov. 14, 1982, submitted for publication in monograph of meeting.

23. "Surface Structure and Properties of R. F. Plasma Polymerized Hexamethyl Disilazane," S. K. Varshney and C. L. Beatty, Organic Coatings and Applied Polymer Sci., 47, Sept. 1982.

24. "Compatibility of Polystyrene-Polystyrene-co-n-butyl methacrylate, Poly(n-butyl methacrylate)-Poly(styrene-co-n-butyl methacrylate) and Poly(styrene-co-n-butyl methacrylate)-Poly(styrene-co-n-butyl methacrylate) Systems," K. Fujioka, N. Noethiger, Y. Baba and C. L. Beatty, National Technical Conference, Society of Plastics Engineers, Symposium on Polymer Alloys, Blends and Composites, Oct. 25, 1982.

25. "Characterization of Polysilastyrene," K. S. Shih, C. L. Beatty, and R. West, to be presented at the High Polymer Div., Am. Phys. Soc. Meeting in Los Angeles, March 21-25, 1983.

26. "Pyrolysis of Polysilastyrene Thin Films as Studied by XPS and IR," K. Shih, T. Saitta, C. L. Beatty, C. D. Batich and R. West, to be presented at the High Polymer Div. of the Am. Phys. Soc. Meeting in Los Angeles, March 21-25, 1983.

27. "Integranular Segregation of Boron in Sintered Silicon Carbide," W. D. Carter, P. H. Holloway, C. White and R. Clausing, submitted to J. Amer. Ceram. Soc., 1983.

28. Y.-X. Wang, F. Ohuchi and P. H. Holloway, "Mechanisms of Electron Stimulated Desorption from Soda-Silica Glass Surfaces," submitted to Surf. Interface Anal., 1983.

29. "Electron Beam Effects during Analysis of Glass Surfaces," P. H. Holloway, presented at Max Planck Institute fur Metallsforschung, Stuttgart, May 24, 1982.

30. "Mechanism of ESD from Glass Surfaces," P. H. Holloway, Florida Chapter of American Vacuum Society, 12th Annual Symp., Clearwater Beach, FL, Feb. 16, 1983.

31. "Electron Stimulated Desorption: Application and Mechanisms," P. H. Holloway, Pittsburgh Conference on Analtyical Chemistry and Applied Spectroscopy, Atlantic City, NJ, March 10, 1983.

32. "Electron Beam Effects during Glass Analysis," P. H. Holloway, National Bureau of Standards, Gaithersburg, MD, March 7, 1983.